ASAE Publication 12-82

ADVANCES IN DRAINAGE

Proceedings of the Fourth National Drainage Symposium

December 13-14, 1982
Palmer House
Chicago, Illinois

Published by
American Society of Agricultural Engineers
2950 Niles Road • St. Joseph, Michigan 49085
(616) 429-0300

Library of Congress Card Number (LCCN) 82-73876
International Standard Book Number (ISBN) 0-916150-47-X

Table of Contents

The 4th National Drainage Symposium is a specialty conference in the continuing tradition of the American Society of Agricultural Engineers of promoting communication between science, industry and practice. The symposium is designed to bring together engineers, scientists and contractors with a common interest in agricultural drainage. Previous drainage symposia were held in 1965, 1971 and 1976. The earlier symposia were developed around specific themes: Drainage for Efficient Crop Production, Methods and Materials for Subsurface Drainage, and Drainage for Increased Crop Production and a Quality Environment. In 1965, the emphasis was on drainage requirements of plants, characterizing soil properties, new methods and materials (plastic drain tubing had appeared to assume a place on the program with clay and concrete pipe), and drain design criteria. In 1971, drain envelopes received considerable attention along with drain pipes and high speed installation techniques. The rapid acceptance of new materials brought a need for a session on new machines and installation methods. At the 3rd National Drainage Symposium in 1976, the program had to be expanded to concurrent sessions and included General Drainage, Evaluation of Drain Envelopes, Models for Drainage Design and Evaluation, New Materials and Installation Methods for Drain Tubes, Drainage for Salinity and Water Quality Control, Physical Properties of Soils Related to Drainage, and Drainage Requirements for Crop Growth. The nature and emphasis of the programs and papers presented in each symposium have reflected the current interest in drainage research and practice.

The 4th National Drainage Symposium is also structured to treat current problems and interests. The planning committee decided on three main topics, Drainage Design, Construction Techniques and Materials, and Drainage of Heavy Soils. The first two topics have been included in some form in previous symposia. The third topic is new in that drainage of heavy soils is now a specific problem of international as well as national interest. Half of the papers on the Drainage of Heavy Soils program are from other countries.

Drainage of agricultural lands is one of the keys to efficient crop production. Use of crop modeling techniques combined with drainage design procedures is becoming a standard practice that will improve the productive and economic efficiency of modern agriculture. These new developments are so well accepted that they do not require a special place on the program, but are part of other papers.

The 4th National Drainage Symposium will not be the last one. There are many problems remaining to be solved in the theory, application, and construction of drains. Acceptance of corrugated plastic drain tubing and laser grade control by farmers and drainage contractors was very rapid in the recent past. Progress in drainage technology in the future may not be as spectacular from a physical standpoint but it will be equally significant. Computers, large and small, are already being used in all phases of drainage and promise better designs for the future. Remote sensing may also become important in the solution of drainage problems. The future may be even more exciting than the past.

A drainage symposium is a team effort requiring voluntary cooperation of many individuals over a long period of time. Some of those involved in the 4th National Drainage Symposium have also contributed to the success of the other three symposia. The committees have worked efficiently to produce an outstanding meeting. As the idea surfaced that it was time for another drainage symposium, an informal group got together at an ASAE National Meeting to do some preliminary planning. William W. Donnan, Glenn O. Schwab, James L. Fouss, Walter D. Lembke, Jan vanSchilfgaarde, R. Wayne

Skaggs, Lyman S. Willardson, Walter J. Ochs, Robert J. Alpers and R.C. Reeve were present in one of the early planning meetings. In that and subsequent meetings a chairman, vice chairman, and committee heads were selected. The general form and subjects for the program were decided and the symposium was under way. The committees were:

Publicity - Melville L. Palmer, Ohio State University,
    Chairman, W.E. Altermatt, C.J. Drablos, J.R. Johnston

Program - Darrell W. DeBoer, South Dakota State University,
    Chairman, W.E. Altermatt, C.E. Carter, W.W. Donnan,
    C.J. Drablos, B.E. Easton, R.W. Irwin, W.R. Johnston,
    D.B. Palmer, R.W. Skaggs, R.D. Wenberg, R.J. Winger

Finances - Ronald C. Reeve, Advanced Drainage Systems,
    Chairman, G.O. Schwab, M.A. Purschwitz

Local Arrangements - James F. Gastel, J.F. Gastel and Assoc.
    Cons. Eng., Chairman, F.H. Galehouse, P.N. Walker

Evening Program - Byron H. Nolte, Ohio State University,
    Chairman, N.R. Fausey, R.J. Alpers

Proceedings - George J. Kriz, North Carolina State University

Besides those officially listed on committees, many other members of the Society have given assistance. The fact that their contributions are not mentioned specifically does not diminish the importance of their ideas and help. Special thanks should be given to Dr. George J. Kriz for his skillful handling of all the proceedings manuscripts.

Other societies and interested organizations were asked to co-sponsor the Symposium. They include:

American Society of Agronomy
American Society of Civil Engineers
Canadian Society of Agricultural Engineering
Corrugated Plastic Tubing Association
Crop Science Society of America
U.S. Committee, International Commission on Irrigation,
    Drainage and Flood Control
Soil Conservation Society of America
Soil Science Society of America
Society of American Foresters
Land Improvement Contractors of America

These organizations appointed official representatives and made possible wide dissemination of symposium announcements. The Corrugated Plastic Tubing Association provided an enabling grant that made participation by overseas scientists possible.

We give our sincere thanks to all who have assisted in making the 4th National Drainage Symposium a success and join in looking forward to the 5th National Drainage Symposium in 1987.

Lyman S. Willardson, Chairman
Walter J. Ochs, Vice-Chairman

A TRIBUTE TO THE AUTHORS

The philosophy of the Steering Committee for the 4th National Drainage Symposium was to have a Proceedings available during the Symposium. In order to adhere to this philosophy, rigid guidelines and dates were established and distributed to the participants. I wish to thank all participants in meeting these deadlines.

The American Society of Agricultural Engineers has the policy that contributions to Proceedings are non-refereed. However, in order to maintain some level of quality control, the contributing papers were sent to two participants to be critiqued. I wish to thank the participants for their cooperation in assisting me in preparing a quality product.

George J. Kriz,
Editor

JAMES N. LUTHIN

Drainage Science lost one of its greatest leaders and scholars in the death of Dr. James N. Luthin on December 24, 1981, following a brief illness. He was Professor of Water Science and Agricultural Engineering at the University of California, Davis, from 1949 until his death at age 66. During the last three decades Dr. Luthin established himself as a world leader in drainage engineering, groundwater hydrology and soil moisture studies. He was responsible for developing his University's irrigation and drainage research laboratory and building the largest tank model in the United States for subsurface drainage and salinity studies. He applied drainage theory to the solution of field problems, especially those involving salinity control for irrigated lands. Since 1972, his research has included coupled heat and water transfer in arctic and subarctic soils. He did collaborative research at the Institute of Water Research at the University of Alaska in Fairbanks and the Institute of Low Temperature Science at Sapporo, Japan.

Dr. Luthin was a Fellow of the American Society of Agricultural Engineers, the American Society of Agronomy and Soil Science Society of America. He was particularly active in the American Society of Agricultural Engineers and served as Chairman of the Society's Committee on Drainage of Irrigated Lands. He was also active in the American Society of Civil Engineers and was serving as the chairman elect of the Executive Committee for the Society's Irrigation and Drainage Division. He received the Hancock Soil and Water Engineering Award from the American Society of Agricultural Engineering in 1969 and was inducted into Ohio State University's "Drainage Hall of Fame" in 1981.

To those of us who knew him, Professor Luthin was an inspiration. He dared to challenge and to lead. His contributions to drainage technology and science will live for many years.

Prepared by George S. Taylor
August 26, 1982

INCREASED RESOURCE PRODUCTIVITY THROUGH DRAINAGE
Jan van Schilfgaarde
Fellow ASAE

It is an honor, as well as a pleasure, to be invited to this lectern for
this, the fourth National Drainage Symposium. I was chairman for the first
of these conferences and participated in the other two, as well. In fact,
you may be getting tired of seeing me up here, but I am pleased to be part
of what is taking on the magic of tradition.

Since the first Conference in 1965, there has been good progress in drainage even though drainage has had a bad press. To word it differently,
there have been significant breakthroughs in research, and substantial
changes in the industry; the industry has served the farmer well. Yet
there has been deemphasis in terms of resources devoted to drainage
research and there continues to be concern voiced in some circles that
drainage is bad, antisocial, counter to man's interest.

Whatever the ever-changing projections for world population growth, whatever the absolute quantitative demands for food in the future, and in spite
of the current economic hardships encountered in U.S. agriculture, all
signs point to increased demand for products from the land, to widespread
food shortages in various parts of the world and, let us not forget, to
continued dependence on agriculture for a substantial portion of our
foreign exchange earnings.

Water management, with drainage as an integral part, is crucial to maintaining or increasing agricultural production. Water management, and thus
drainage, are important components of a strategy towards resource conservation. Let's hope that, besides the successes in research and development,
drainage will lose its undeserved stigma and receive the recognition it
deserves in helping alleviate the pending water and production problems.

PROGRESS IN THE ART OF DRAINAGE

In the few minutes available, a balanced review of drainage activities is
not possible, and the occasion doesn't call for it. Instead, I choose to
high-light a few events that, to me, accent the progress made, more or less
since the first of our Conferences.

Design

In 1965, two papers on design told how design criteria were generally based
on an accumulation of field experience. Ron Reeve even suggested one possible definition of design as: "try something and modify as needed on the
basis of performance". Theories had been developed, and some of us had
made as yet unsuccessful attempts to introduce the use of such theories
into rational design procedures for practicing engineers. However, we
didn't have confidence in methods to characterize field parameters; we

---

The author is: JAN van SCHILFGAARDE, Director, U. S. Salinity Laboratory,
USDA, Riverside, CA.

didn't have adequate data to relate design criteria to crop performance; we couldn't massage the volume of data well enough to handle the probability aspects we thought should be used. When the ASA Drainage Monograph was published in 1974, the situation was not very different. In fact, I myself made the plea at least twice in that tome that an effective linkage be forged between rational design procedures and practicing engineers. Wiser, Ward and Link (1974) had extended design concepts to take into account crop response and financial return, but the data base was still lacking for practical use. Long-time field work by Schwab et al. (1975) in Ohio, and crop response studies by Hiler (1977) and students in Texas were beginning to provide the data base on crop response that was needed. The extensive work of Skaggs (1978) led to the now well-known DRAINMOD. Now, in 1982, we see evidence of a pregnant union between crop response data and hydraulic models, of successful coordination between researcher and field engineer. I, for one, am greatly encouraged by Skaggs' imput to the SCS Engineering Handbook, facilitated by Walter Ochs, and by the evidence that the efforts of so many, especially in Ohio, Texas, and North Carolina, have paid off (e.g., Hardjoamidjojo + Skaggs, 1982).

I realize we still have a long way to go. I recognize that the data base still is weak. I suspect that routine design by computer program is likely to pose some problems in a number of field situations. I am keenly aware that, under irrigated conditions, deep seepage and lateral flow phenomena put severe restrictions on the use of those models that are available to date. Still the progress has been impressive and I salute those among you who have made it happen.

Materials

At the first Conference, Alpers and Short (1965) discussed advances in the clay tile industry; the Proceedings carried a set of photographs of pallets with clay tile. Not long after, Alpers' company discontinued making clay tile and switched to corrugated tubing. That observation dramatically illustrates the changes that have taken place. The switch to plastic tubing has been almost total. Standards have been worked out. Government specs are in place and few arguments are heard today on the choice between tubing and clay or concrete tile.

In the related area of envelope materials, there also has been considerable activity, but in my opinion, not quite as much progress. Since Kirkham's (1950) early work on what we now call entrance resistance, a lot of water has flowed through the cracks and perforations, and some has been held back by inoperable filters or chemically clogged envelopes. We have tried and discarded numerous filter fabrics, learned how to use others with reasonable success. We have discovered that chemical clogging of drains is much more widespread than most of us realized; and we have learned a good deal about the process, but as yet do not know how to predict or prevent its occurrence (Ford, 1979). Willardson and Walker (1979) have proposed a critical hydraulic gradient to assess the need for (and effectiveness of) drain envelopes. At a drainage conference in the Netherlands (Wesseling, 1978), extensive discussion indicated the substantial effort, here and abroad, that has gone into the development of elaborate theories describing entrance resistance and the testing of numerous envelope materials, both in the laboratory and in the field. In all engineering process situations, there is a point of greatest weakness; I suspect that, in drainage, that point at present is in selection of the most appropriate envelope.

Installation

As with drain materials, there has been a drastic upheaval in installation techniques. It is interesting to reread the papers by Fouss (1964) and Ede (1964) in the first Proceedings. Many of the principles incorporated in today's plows and trenchers were discussed at that time, but commercial application is certainly of more recent origin. Installation speeds have increased at least tenfold. Laser grade control is now commonplace. Trenchless installation now goes side-by-side with trenching.

As experience has been gained, some problems have arisen that not everyone anticipated. With the high speeds, quick response for grade control on trenchless plows is crucial and not always obtained. Inspection or feed-back techniques to verify proper installation or identify trouble spots are still lacking. Soil compaction when installations are made beyond the critical depth, or in wet, fine textured soils, is a problem that needs further attention.

FUTURE TRENDS

Predicting the future is not my forte; it is best left to seers and econ-omists. It is more to the point to identify some of the problems that face us, in the hope that some, at least, will be resolved in the near future.

First, I see a continuing and increasing recognition of the need to con-sider total water management. In humid areas, total water management may include sub-irrigation as well as subsurface drainage, surface drainage or land leveling; it may also take into account the possibility of over-drainage, and means of preventing it. The concept isn't new, nor its application, but the fuller realization of the potential benefits from water management may well lead to more sophisticated and better integrated technologies in the future than we have used traditionally. In arid areas, the interdependence of drainage and irrigation should be blatantly obvious. I shall return to that point later.

Closely associated with integrated water management is the (ever present) need for better data. Though the data base that has been used in recent efforts at rationalizing drainage design has served an excellent purpose, additional data on crop response and trafficability would greatly enhance the field use of the new design procedures. Similarly, there continues to be a need for more effective and reliable means to obtain appropriate quantitative numbers for soil properties.

Drainage of irrigated lands introduces some other problems that often are not explicitly recognized. First, the drainage intensity (or the rate at which water must be removed) clearly depends very strongly on the irriga-tion management; for example, a change in on-farm irrigation efficiency from 80% to 90% halves the amount of drainage required. Second, the amount of drainage needed is not strictly determined by deep percolation from cropped fields; natural drainage may be sufficient without man's assist, or conversely, canal seepage may completely dominate the water balance. Obvious as these observations may be, the drainage design standards with which I am familiar do not take these factors into account; and when Wayne Skaggs and Andrew Cheng recently analyzed a fairly extensive data set in the context of DRAINMOD, they concluded that the deep seepage component was crucial in obtaining "closure". In the case of irrigated lands, "total water management" is concerned not only with the water on the land, but also with that in the distribution system. Furthermore, water quality problems enter in far more prominently than they do in humid areas. In fact a main objective of drainage in irrigated lands is salinity manage-ment. Having just stated that design criteria for these circumstances require improvement, I need to add that substantial progress has been made in establishing the drainage requirement for salinity control, in the sense of a leaching requirement.

With reference to installation techniques, I expect to see some new developments that will permit trenchless installations at greater depths

3

without the complications of compaction. One pet project of mine so far
has not come to fruition. We need a soil stabilizer, or an artificial
gravel made in place, to substitute for the gravel used in envelopes.
Though several attempts along that line have been made in Belgium, in
Israel and in the U.S., it is my understanding that none was successful.
The only product with which I was associated personally, was water soluble!
Maybe someone sitting in this room will surprise us and find a good solution. Whether or not my artificial gravel makes sense, I am convinced that
flow in the area immediately surrounding the drain still poses a number of
unresolved questions that are important in attaining effective drain
operation.

Another area of concern in drainage theory and practice is the drainage of
fine-textured soils. It is encouraging to note the heavy emphasis on this
subject in the program for this meeting. Though we should anticipate that
the discussion the next two days will shed substantial light on the subject, we may be sure that not all problems will have been solved. To what
extent can we predict changes in soil structure and in hydraulic conductivity following drainage and intensive cropping? With renewed interest
across the country in controlled trafic, or zone tillage, as a farming
practice to avoid soil compaction, will new opportunities arise for
improved water management?

THE CHALLENGE

Drainage is the subject of this Conference, but drainage does not exist
in isolation. It constitutes one component of a water management system
that, in turn, makes up a component of a crop production system. In many
instances, water mangement--and with it drainage--is a dominant component
of a productive system of agriculture.

Without irrigation, crop production in Egypt would be close to nil. With
irrigation, production is reasonably high. With drainage as well, it can
be outstanding. Water logging is a major constraint to production in
Pakistan as well. But why go so far from home? The Central Valley in
California demonstrates the same relations. The Westlands Water District
near Fresno has a serious drainage problem; the only outlet is an evaporation pond and it is full. My colleagues started a research project four
years ago in an area of newly developed land, with imported water. At the
time, the water table in the region was at least 5 m deep. Last year and
this, it came up to 40 cm. The 30-year old warnings that drainage is a
necessity if agriculture is to be sustained have taken on new urgency.

Seconding frequent comments by Glenn Schwab, wetland drainage is not a
sin, is not exploitation; it is a constructive technique for increasing
sustained production from natural resoures, and thus a resource conservation measure.

Those of us who are interested in soil and water conservation and who have
devoted substantial effort to improving the art and science of drainage,
know that good water management contributes substantially to better
resource use--to energy conservation, to reductions of erosion, to
increased water use efficiency and, of course, to increased crop production
and reductions in production costs. We know that, without adequate water
management, crop production would suffer drastically.

Thus we take pride in the accomplishments of the past and look forward to
continuing contributions in the future, contributions to resource conservation, to food production and thus to human well-being. At this Conference,
many of the leaders in drainage are gathered together. I salute you and
praise you for your past endeavors; I wish you success into the future.

REFERENCES

1. Alpers, R. J., and T. N. Short. 1965. Clay draintile and its use. ASAE Conf. Proc. Drainage for Efficient Crop Production 46-48.

2. Benz, L. C., E. J. Doering, and G. A. Reichman. 1981. Watertable management saves water and energy. TRANSACTIONS of the ASAE 24(1):995-1001.

3. Ede, A. N. and G. M. Smitham. 1965. New materials and machines for drainage. ASAE Conf. Proc. Drainage for Efficient Crop Production 58-60, 68.

4. Ford, Harry W. 1979. Characteristics of slime and ochre in drainage and irrigation systems. TRANSACTIONS of the ASAE 22(5):1093-1096.

5. Fouss, James L. 1965. Plastic drains and their installation. ASAE Conf. Proc. Drainage for Efficient Crop Production 55-57.

6. Hardjoamidjojo, S., and R. W. Skaggs. 1982. Predicting the effects of drainage systems on corn yields. Agric. Water Manage. 5(2):127-144.

7. Hiler, Edward, A. 1977. Drainage requirements of crops. Proc. Third National Drainage Symposium. ASAE Pub. 1-77:127-129.

8. Kirkham, D. 1950. Potential flow into circumferential openings in drain tubes. J. of Applied Physics 21:655-666.

9. Schwab, G. O., N. R. Fausey, and C. R. Weaver. 1975. Tile and surface drainage of clay soils. II. Hydrologic performance with field crops (1962-1972). III. Corn, oats, and soybean yields (1962-1972). Res. Bull. 1081, OH Agric. Exp. Wooster, OH, 237 pp.

10. Skaggs, R. W. 1978. A water management model for shallow water table soils. Rep #134 Water Resour. Res. Inst., Univ. North Carolina, 27650. 178 pp.

11. Wesseling, J. Proceedings of the International Drainage Workshop. 1978. J. Wesseling (ed.) ILRI Pub. 25. 731 pp.

12. Willardson, Lyman S., and Robert E. Walker. 1979. Synthetic drain envelope--Soil interactions. J. Irrigation and Drainage Division, ASCE. 105(IR4:367-373.

PREDICTING WATER TABLE DRAWDOWN FOR TWO-DIMENSIONAL DRAINAGE

R. LAGACE   R. W. SKAGGS   J. E. PARSONS
                         Member ASAE

Water table movement due to drainage ditches or tubes can be described by solving the Boussinesq equation which is based on the Dupuit-Forchheimer (D-F) assumptions. The one-dimensional case for water table drawdown between parallel drains has been investigated thoroughly and a number of solutions are known. The two-dimensional case, such as exists in the area where parallel drains intersect a main, has only been investigated for a few specific cases and solutions generally are not available. One method of approximating the solution to the two-dimensional case is to linearize the Boussinesq equation. Then the variables are separable and the solution can be obtained as the product of the two one-dimensional solutions as applied for certain transient two-dimensional heat transfer problems. The purpose of this paper is to determine the validity of using the product of the one-dimensional solutions to the nonlinear Boussinesq equations as the solution for the two-dimensional case.

THEORY

Assuming the validity of the D-F assumptions, the water table behavior in time and space can be described by the nonlinear Boussinesq equation,

$$f \frac{dh}{dt} = \frac{\partial}{\partial x} \left(K_x \, h\frac{\partial h}{\partial x} \right) + \frac{\partial}{\partial y} \left(K_y \, h\frac{\partial h}{\partial y} \right) + R \tag{1}$$

h = water table height above the impermeable layer (Fig. 1)
K_x, K_y = hydraulic conductivity in x and y directions
f = drainable porosity or specific yield
t = time
R = rate of rainfall, evapotranspiration or deep seepage.

For the linearized form of Eq. (1) with R = 0, Carslaw and Jaeger (1959) have shown that if the problem can be transformed to a 0 and 1 boundary condition problem and if the solution is continuous along the boundaries, the solution to the two-dimensional differential equation is the product of the solutions of the differential equations for each direction:

$$\phi(x,y,t) = \phi(x,t)\,\phi(y,t) \tag{2}$$

where $\phi(x,y,t)$ is the solution for transient two-dimensional flow, and $\phi(x,t)$ and $\phi(y,t)$ are solutions for transient flow in the x and y directions, respectively. This approach considerably simplifies the solution of problems in two or three dimensions by use of the known solutions for the corresponding one-dimensional cases.

The authors are: R. LAGACE, Graduate Student, R. W. SkAGGS, Professor, and J. E. PARSONS, Graduate Student, Biological and Agricultural Engineering Dept., North Carolina State University.

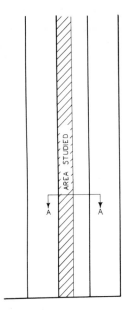

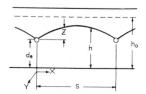

Fig. 1   Schematic of the Soil
         Profile.

a) Parallel drains flowing into a   b)   Square grid system.
   main (Gridiron system).

Fig. 2   Two-Dimensional Drainage Problems Studied

For nonlinear differential equations, this priciple does not hold mathemati-
cally. But because the Boussinesq equation is not highly nonlinear, our
hypothesis is that this principle will yield an approximate solution with
acceptable errors. Because many agricultural drainage problems can be expressed in terms of 0 and 1 boundary conditions, this approximation could be
very useful for both practitioners and researchers.

The problems considered herein are for two-dimensional drainage with an
initially horizontal water table. Analytical solutions to the Boussinesq
equation are not available for either one- or two-dimensional flow for these
initial conditions. Some approximate solutions have been published for the
one-dimensional case of parallel drains and for the case of a single drain
in a semi-infinite medium. They can be grouped as linearized form solutions
(Dumm, 1954; Dumm, 1964; Kraijenhoff van de Leur, 1958; Massland, 1959),
numerical solutions presented in graphical form (Moody, 1966; Skaggs, 1975,
1979) and analytical solutions for the parabolic initial water table shape
(van Schilfgaarde, 1963). For the special case of the well developed regime,
a more exact solution based on the theory of potential flow can be used
(Kirkham, 1964; Guyon, 1966, 1980).

7

For the two-dimensional case, the only practical way to obtain a solution of the differential equation is by the use of numerical methods. We chose the finite element method to describe the problem in space and a semi-implicit finite difference scheme for the time axis. The finite element method permits very fine gridding in the region of fast changes (close to the drains) and coarse gridding in regions of slow changes farther removed from the drains. This reduces computation time. This method was used for the Boussinesq equation by Tzimopoulos (1976). The elements used are rectangular with eight nodes, located at each corner and in the middle of each side, and triangular elements with six nodes. Both are quadratic in their approximation functions and are very reliable (Zienkiewick and Parekh, 1970). The finite element program was developed by Dhatt and Touzot (1981).

In order to preserve generality, we worked with the non-dimensional form of the Boussinesq equation which may be written for R = 0 as,

$$\frac{dH}{d\tau} = \frac{\partial}{\partial \xi} \left(H \frac{\partial H}{\partial \xi} \right) + \frac{\partial}{\partial \eta} \left(H \frac{\partial H}{\partial \eta} \right). \tag{3}$$

where $H = h/h_o$
  $\xi = x/S$
  $\eta = y/S$
  $\tau = \dfrac{K h_o t}{fS^2}$
  h = initial water table height above the impermeable layer
  S = spacing between drains.

The non-dimensional time axis was divided into 48 intervals of τ increasing from 0.00001 to 0.05. To minimize the possibility that the error caused by different numerical schemes would mask the error due to the difference between the two-dimensional solution and the product of two one-dimensional solutions, the same numerical scheme was used for the one-dimensional and two-dimensional cases (same time steps with the same gridding on equivalent axis). For the one-dimensional case, our solution was compared to the numerical solution presented by Skaggs (1975) using a finite difference method with a very fine grid and very small time intervals. Our solution was always within 1% of his solutions for both the parallel drains and single drain cases.

Two cases were analyzed. The first deals with the very common gridiron system where parallel ditches or drains flow to a main (Fig. 2a). The second deals with a square grid of ditches or drains (Fig. 2b). The second case is less common in practice but more interesting in mathematical terms because each drain influences all points within a given square. Such a network has been proposed for draining very heavy soils. In the remaining text we will mention only the terms "drain" with the meaning of ditch or drain. If drain tubes are used, procedures such as the use of Hooghoudt's equivalent depth (d_e) (van Schilfgaarde, 1974) to correct for convergence near the drain must be included.

GRIDIRON SYSTEM

In terms of one-dimension solutions, the gridiron system is equivalent to the combination of parallel drains with a single drain (semi-infinite case). Because of the symmetry of the system, only half of the area between two drains is analyzed. The meshing of the gridiron is presented in Fig. 3. We limited our investigation to five times the drain spacing (5S) in the direction perpendicular to the main. As will be shown later, the main does not influence water table drawdown beyond a distance of less than two drain spacings (2S) so our limitation of 5S did not affect the results.

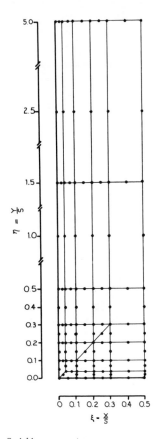

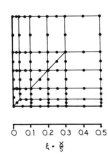

a) Gridiron system   b) Square grid system

Fig. 3   Finite Element Meshing for the Gridiron and
Square Grid Drainage System

If d_e is the equivalent depth from the drain to the impermeable layer, the
boundary conditions for the gridiron system can be written as,

1. $H = 1$, $0 \leq \xi \leq 0.5$, $0 \leq \eta \leq 5$, $t = 0$ (4a)

2. $H = d_e/h_o$, $\xi = 0$, $0 \leq \eta \leq 5$, $t > 0$ (4b)

3. $\dfrac{\partial H}{\partial \xi} = 0$, $\xi = 0.5$, $0 \leq \eta \leq 5$, $t > 0$ (4c)

4. $H = d_e/h_o = D$, $0 \leq \xi \leq 0.5$, $\eta = 0$, $t > 0$ (4d)

5. $\dfrac{\partial H}{\partial \eta} = 0$, $0 \leq \xi \leq 0.5$, $\eta = 5$, $t > 0$ (4e)

All boundary conditions were applied to obtain the solution for the two-
dimensional case. The one-dimensional solutions were obtained by applying
conditions 4a, 4b, and 4c for flow in the x-direction and conditions 4a, 4d,
and 4e for flow in the y-direction. Solutions were determined for three
cases of the gridiron system corresponding to $D = d_e/h_o$ values of 0.1, 0.2
and 0.6. Both the one- and two-dimensional solutions were transformed
according to the equation,

$$Z = z/z_o = (H - D)/(1 - D) = (h - d_e)/(h_o - d_e)$$ (5)

DRAINAGE IN TWO DIMENSIONS
D=d_e/h_o =0.1

Z (x) Z (y)

Z (x , y)

Fig. 5 Comparison Between the Product of Solutions $Z(x)Z(y)$ for One-Dimensional Flow in the x and y Directions with the Solution for Two-Dimensional Flow $Z(x,y)$ for a Gridiron Drainage System with $D = d_e/h_o = 0.1$

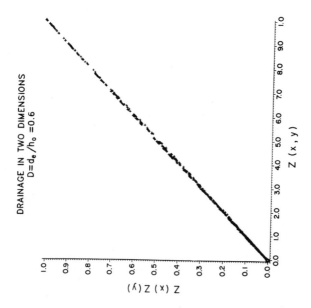

DRAINAGE IN TWO DIMENSIONS
D=d_e/h_o =0.6

Z (x) Z (y)

Z (x , y)

Fig. 4 Comparison Between the Product of Solutions $Z(x)Z(y)$ for One-Dimensional Flow in the x and y Directions with the Solution for Two-Dimensional Flow $Z(x,y)$ for a Gridiron Drainage System with $D = d_e/h_o = 0.6$

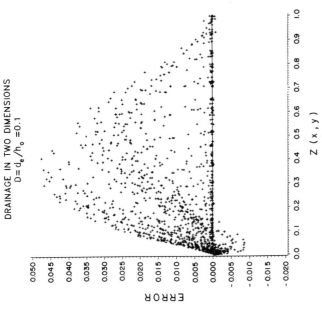

DRAINAGE IN TWO DIMENSIONS
$D = d_e/h_o = 0.1$

Fig. 7 Error of the Product Solution $Z(x)Z(y)$ as Compared to the Two-Dimensional Solution $Z(x,y)$. The Error was Computed as, $ERROR = Z(x,y) - Z(x)Z(y)$. Recall that Z Varies from 0 to 1 so that the Error Computed Represents the Fraction of the Maximum Z Value

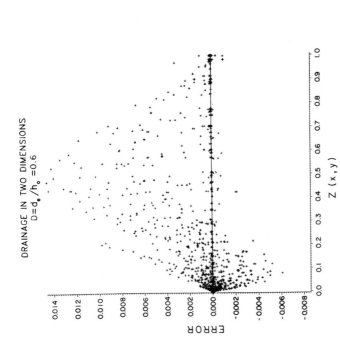

DRAINAGE IN TWO DIMENSIONS
$D = d_e/h_o = 0.6$

Fig. 6 Error Caused by Using the Product Solution $Z(x)Z(y)$ as Compared to the Two-Dimensional Solution $Z(x,y)$. The Error was Computed as, $ERROR = Z(x,y) - Z(x)Z(y)$. Recall that Z Varies Between 0 and 1 so that the Error Computed Represents the Fraction of the Maximum Z Value

where $z_o = h_o - d_e$ is the initial water table elevation above the drain. This transformation forces the boundary conditions to conform to the zero-one requirement. Z actually represents the nondimensional elevation of the water table above the center of the drain, which varies from 0 to 1.

Figures 4 and 5 show comparisons between the product solution and the two-dimensional solution for D = 0.6 and D = 0.1, respectively. $Z(x,y)$ is the solution to the two-dimensional equation, while $Z(x)$ and $Z(y)$ represent the solutions for one-dimensional flow in the x and y directions, respectively. The correlation was good for both D values ($R^2 > 0.98$), but was better for D = 0.6 because there was less variation in H and the equation was more linear than for D = 0.1. Figures 6 and 7 show the error structure. The error was computed as, Error = $Z(x,y) - Z(x)Z(y)$. Because Z varies from 0 to 1 an error of 0.015 represents 1.5% of the maximum Z value of the water table position. The maximum errors of the product solution were 1.5% (D = 0.6), 3.8% (D = 0.2) and 4.8% (D = 0.1), and these occurred when Z was approximately 0.3 for all D values. The maximum error also occurred at a point midway between drains ($\xi = 0.5$ or x = S/2) and at an equal distance from the single drain in the y direction ($\eta = 0.5$ or y = S/2).

As a check of the numerical methods, the same meshing with the same number of time steps was run for the linear differential equation and the maximum error was found to be 0.5%. This represent the portion of the error due to the numerical approximation.

SQUARE GRID SYSTEM

In terms of one-dimensional solutions, the square grid system is equivalent to a combination of two parallel drainage systems. The meshing used is presented in Fig. 2b. For this system, only two cases were analyzed: D = 0.2, and D = 0.6.

Figure 8 presents a comparison between the product solution $Z(x)Z(y)$ and the two-dimensional solution $Z(x,y)$ for the midpoint ($\xi = 0.5$, $\eta = 0.5$). Results for the same point on the gridiron system also are plotted for comparison. The results show that the product was in good agreement with the two-dimensional solution for both D values. The error for the square grid system was somewhat larger than for the gridiron system as shown more clearly in Fig. 9 where the error at the midpoint is plotted as a function of Z at that point. The maximum error occurs at the midpoint and was 7% for D = 0.2 and 2.7% for D = 0.6.

APPLICATION

An important question for researchers and practitioners involves the area or distance of influence of the main drain. We can use both methods, the two-dimensional solution and the product solution, to approximate this area of influence. Results of an analysis for a range of tolerable errors are presented in Table 1. The influenced areas predicted by the product solutions are larger than those predicted by the two-dimensional solutions for the errors tested. From a practical point of view, the distance predicted by the product solution is larger and can be used as a safe estimate of the area of influence of the main. From these results we can conclude that the main would have practically no influence on water table drawdown at distances larger than two drain spacings removed from the main. It should be emphasized that these results hold only for cases where the main drain is at the same depth as the laterals.

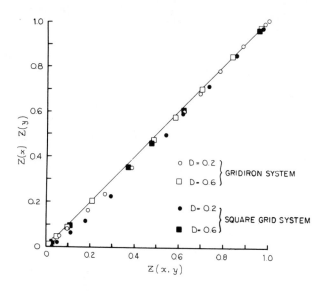

Fig. 8 Product Solutions $Z(x)Z(y)$ for the Point $\xi = X/E = 0.5$,
$\eta = Y/E = 0.5$ as Compared to Two-Dimensional Solutions for
$D = 0.2$ and $D = 0.6$ for Both the Gridiron and Square Grid
Drainage Systems

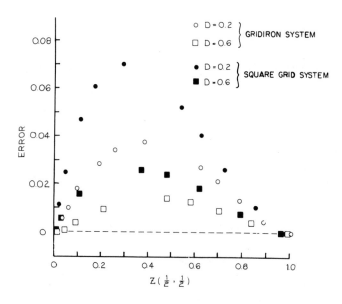

Fig. 9 Error of the Product Solution at a Point Midway Between Drains
$(\xi = 0.5, \eta = 0.5)$ for Both the Gridiron and Square Grid Systems.
Note the Maximum Error Occurs when $Z = 0.3$

Table 1. Distance From Which the Main Drain Influences the Water Table
          (In Terms of Drain Spacing)

Error [*]	Two-dimension solution D			Product solution D		
	0.1	0.2	0.6	0.1	0.2	0.6
1%	1.1	1.15	1.2	1.9	1.6	1.3
2%	0.84	0.90	0.96	1.4	1.3	1.1
5%	0.60	0.62	0.74	0.87	0.86	0.80

[*] Maximum error in terms of the initial head

SUMMARY AND CONCLUSIONS

The product of two one-dimensional solutions were used to approximate two-dimensional solutions for drainage problems described by the nonlinear Boussinesq equation. Two cases were evaluated. The first case involved parallel drains which intersect a main (gridiron). The correlation between the product of one-dimensional solutions and the two-dimensional solution was 0.98. The maximum errors in the product solutions ranged from 1.5% to 4.8% with an estimated error of 0.5% due to the numerical procedure. The second case tested involved a square grid of drains. For this case, the maximum error ranged from 2.7% to 7%.

In both cases, the error was within a tolerable range for engineering design. The one-dimensional solutions may be obtained from graphical plots or approximate equations. Therefore the product solutions are much easier to compute and apply than numerical solutions to the two-dimensional equations.

REFERENCES

1. Carslaw, H. S. and J. C. Jaeger. 1959. Conduction of Heat in Solids. Oxford University Press, N.Y.

2. Dhatt, G. and G. Touzot. 1981. Une présentation de la méthode des éléments finis. Les Presses de l'Université Laval, Québec.

3. Dumm, L. D. 1954. Drain spacing formula. Agricultural Engineering 35:726-730.

4. Dumm, L. D. 1964. Transient-flow concept in subsurface drainage: Its validity and use. TRANS. of the ASAE 7(2):182-186.

5. Gureghian, A. B. 1978. Solution of Boussinesq's equation for seepage flow. Water Resources Research 14(2):231-236.

6. Guyon, G. 1966. Considération sur l'hydraulique du drainage des nappes. Bull. Tech. de Génie Rural, No. 79. C.T.G.R.E.F., Antony, France.

7. Guyon, G. 1980. Transient state equations of water table recession in heterogeneous and anisotropic soils. TRANS. of ASAE 23(3):653-656.

8. Kirkham, D. 1964. Physical artifices and formulas for approximating water table fall in tile-drained land. Soil Sci. Soc. Amer., Proc. 28:585-590.

9. Kraijenhoff van de Leur, D. A. 1958. A study of non-steady ground-water flow with special reference to a reservoir-coefficient. De Ingenieur 40:87-94.

10. Maasland, M. 1959. Water table fluctuations induced by intermittent recharge. J. Gerphys. Res. 64:549-559.

11. Moody, W. T. 1966. Nonlinear differential equation of drain spacing. J. Irr. and Drain. Div., ASCE, 92(IR2):1-9.

12. Skaggs, R. W. 1975. Drawdown solutions for simultaneous drainage and ET. J. Irr. and Drain. Div., ASCE, 101(IR4):279-291.

13. Skaggs, R. W. 1979. Factors affecting hydraulic conductivity determinations from drawdown measurements. ASAE Paper No. 79-2075, presented at the 1979 Summer ASAE Meeting, Winnipeg, Canada.

14. Tzimopoulos, C. 1976. Solution de l'équation de Boussinesq par une méthode des éléments finis. J. of Hydrol. 30 1-18.

15. van Schilfgaarde, J. 1963. Design of tile drainage for falling water table. J. Irr. and Drain. Div., ASCE, 89(IR2):1-12.

16. van Schilfgaarde, J. 1974. Nonsteady Flow to Drains. Ch. 11 in Drainage for Agriculture. J. van Schilfgaarde, ed., Amer. Soc. Agron., Madison, WI. pp. 245-270.

17. Zienkiewick, O. C. and C. J. Parekh. 1970. Transient field problem: two-dimensional and three-dimensional analysis of isoparametric finite elements. Int. J. Num. Meth. Eng. 2(1):61-71.

VARIABILITY OF SOIL WATER CHARACTERISTICS OF AN ORGANIC SOIL

J.A. Millette   R.S. Broughton
Assoc. Member ASAE   Member ASAE

Variability of drainage design parameters, frequently unknown for Histosols, hampers economic and efficient subsurface drainage designs. Most of the research on the variability of soil physical properties has been conducted on mineral soils (Nielsen et al. 1973; Warick and Nielsen 1980). Boelter (1964; 1965; 1969; 1974) brought into perspective the hydrological characteristics of organic soils. He found that water retention, porosity, water yield coefficient and hydraulic conductivity varied greatly among different peat types and organic soils. However, within the same peat type there was no report on the variability.

As part of an extensive drainage and water quality experiment, a preliminary investigation was undertaken to determine possible variations of the organic soil physical properties among large drainage lysimeter plots. From initial drain outflow and water table measurements, large variations were noticed, this prompted an investigation to elucidate the problem. If a proper statistical design were to be established for the experiment, spatial variability had to be evaluated among the plots.

The objectives of this investigation were to measure and determine the spatial variability of field measured and calculated soil-water properties in a histosol using drainage lysimeter plots. The measured soil-water properties included water table depths, subsurface drain outflow, hydraulic conductivity and drainable porosity.

MATERIALS AND METHODS

The Histosols of the Farnham Bog in Missisquoi county, Quebec ($45^{\circ}17'N$, $73^{\circ}03'W$) extend over a total area of 2065 ha divided into three lobes. The northern lobe where the experimental site is located has an organic soil depth ranging from 3.5 to 4.5 m. This lobe is classified as a basin bog developed under minerotrophic conditions and is considered as a sapric Borohemist (Lévesque et Millette 1977). The morphological stratigraphical nature of the sedge-peat is difficult to establish in the top 1.5 m because the organic material is well decomposed. The surface material is composed mostly of sedge, wood particles and roots.

Underground shelters (3.8 m x 4.6 m) were constructed during the winter of 1978-79, followed by the construction of twelve lysimeter plots during the summer of 1979. Each 6 x 15 m drainage lysimeter plot was separated by a plastic barrier down to approximately 2 m and is drained by a subsurface and surface drain, both having outlets in the underground shelters (Fig. 1). Corrugated plastic 100 mm diam drain tubes were installed within each plot using a narrow, 300 mm bucket on a backhoe at depths ranging from .90 to 1.20 m and on a grade of 1%. Natural grade from plot 12 to plot 1 was .36%

The authors are: Research Engineer, Research Station, Research Branch, Agriculture Canada, St-Jean-sur-Richelieu, Quebec, J3B 6Z8, Canada, Contribution No. J.908 ; Professor, Agricultural Engineering Dept., Macdonald Campus, McGill University, Ste-Anne-de-Bellevue, Quebec, H0A 1C0, Canada.

(Fig. 1). Thick plywood sheets were placed under equipment tires to prevent sinkage and reduce soil disturbance. Non-perforated 100 mm diam ABS plastic pipe was used to connect the drains from the plot to the underground shelters.

The surface run-off was contained within each plot by a small dike built using organic soil and a plastic drain pipe covered with a black polyethylene plastic sheet, and directed to a surface inlet from which the runoff was measured. Precautions were taken to eliminate outside water from encroaching on the plots by constructing a larger dike surrounding all plots and shelters. Additional subsurface drains were installed around each shelter below the outlet and outside the twelve plots. The purpose of these drains was to maintain the hydraulic head difference at a minimum between the inside and outside of the plots reducing leakage problems.

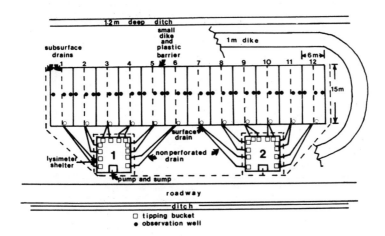

Fig. 1 Schematic Diagram Showing Drainage Lysimeter Plots and Shelters.

Within each plot, three 13 mm diameter observation wells were installed: one as close as possible to the drain; the other two 1.80 m on each side of the subsurface drain (Fig. 1). Water level readings were taken with either a blow tube or an electric probe (Campbell et al. 1980). Water table levels in the observation wells were taken 3 to 5 times per week between May and October 1981.

Water outflow from subsurface drains was measured between December 1980 and December 1981 in the underground shelters with plastic tipping buckets equipped with mechanical counters. The frequency of the readings was the same as for the water table level readings except that they were taken all year long. The volume of water per tip was 400 ml with each double tip counted as one. The tipping buckets were calibrated and showed that outflow discharge rates at 225 l/h could be measured with less than 5% error, and up to 550 l/h with 16% error, when comparing the volume necessary to tip the bucket under a quasi-static situation. The expected maximum outflow calculated from rainfall data is approximately 270 l/h. Most of the flows that occurred were below 100 l/h well within the range where the tipping buckets had high accuracy. Precipitation was measured with a recording rain gauge at the site location from April 15 to November 7, 1981. During the period of study the soil was limed but remained bare in preparation of a future water quality experiment.

Water Table Fluctuations

Precipitation and typical water table levels are shown in Fig. 2 for plot 3. The trend was the same for all the plots. Three main hydrological events occurred during the season: two in August and one in October. Following a dry spell at the end of July, 69 mm of rainfall on August 4 and 5 resulted in high water tables. The second event on August 15 to 17 occurred when 84 mm of rain fell and raised the water table even closer to the soil surface. Finally, the third main hydrological event occurred when 103 mm of rain fell in late October.

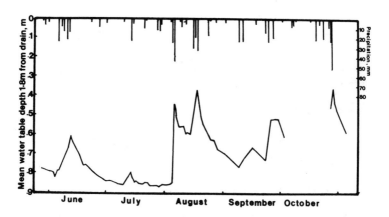

Fig. 2   Water Table Fluctuations and Precipitations for Plot 3.

Water table level comparisons among plots are possible if the water table depth in each plot is subtracted from the drain depth. The water table heights among plots on a particular day were compared to determine the distribution of values and possible groupings. The Kolmogorov-Smirnov one-sample test (Daniel 1978) for normality showed that the water table height values for the 12 plots were most likely normally distributed as also mentioned by Dieleman and Trafford (1976).

Water table heights in the first six plots (lysimeter 1) were compared to those in the last six plots (lysimeter 2). The purpose was to determine whether or not the groups could be separated into two experimental blocks. Certain dates were chosen to test the hypothesis that differences in water table heights between the two lysimeter groups existed. The Mann-Whitney U test was used to avoid the assumptions associated with the t or t' test. The tests showed that the water table heights on some dates were greater in lysimeter group 2 than in the lysimeter group 1, but in other cases there was no significant difference. Probably other factors, such as the antecedent rainfall or hydraulic conductivity (discussed later) could have had an effect.

Outflow Data

Monthly totals of precipitation and drain outflow are shown for plot 3 in Fig. 3. The hydrological events in August and October discussed earlier also appear to be the major events of the 13 month study period. Unusually mild weather resulted in high drain outflow values in February. Drain outflow was almost zero from all drains, in January 1981.

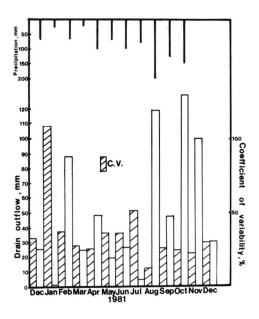

Fig. 3 Monthly Totals of Precipitation and Drain Outflow for Plot 3,
and C.V. of Drain Outflow for All Plots.

The drain outflow coefficient of variability (C.V.) among plots calculated
for each month of the year is shown in Fig. 3. The lowest C.V. of 12.9%
occurred during August whereas the low outflow months of January and July
had C.V. of 108% and 52%, respectively. During the rest of the year the
C.V. ranged from 22.6 to 37.5%. From these data it appears that greater
drain outflow variability among the 12 plots can be expected when outflow
is low than when it is high. Although C.V. values for soil properties
greater than 100% may be symtomatic of skewed distributions (Beckett and
Webster 1971), all drain outflow values were most likely normally distributed
as tested with the Kolmogorov-Smirnov one-sample test. Furthermore, the
variance changes from month to month and is not homogeneous.

To test the hypothesis that the total monthly drain ouflow among the plots
is not different, a Freidman two way analysis of variance was performed
(Daniels 1978). The test does not assume normality nor homogeneous varian-
ces. Furthermore, blocks (plots in this case) can be tested for differences.
The test was highly significant and the conclusion is that at least one plot
is different. Multiple comparisons for use with the Freidman test reveal
that drain outflow from plot 11 is significantly lower than drain outflow
from plots 1, 5, and 12 at the 5% level but not different from the rest of
the plots. Some plots will always yield lower drain outflow rates than
others.

Hydraulic Conductivity

Before calculating the hydraulic conductivity for each plot, drain outflow
values were plotted against hydraulic head. As an example of the type of
relationship, values for plot 3 are shown in Fig. 4. The equation that
best describes the relationship is of the form:

$$q = a\,h^{b} \tag{1}$$

where q = drain outflow
   h = water table height above drain centre or hydraulic head
  a,b = constants determined using the least squares fit.

Of the twelve plots, one could not be used because of incomplete data, and another plot did not show the above relationship. All remaining plots except one had an index of correlation of .904 or greater. The theoretical curve for plot 3 is shown in Fig. 4. The high negative exponent causes the curve to rise rapidly past 300 mm of hydraulic head. Heads below 300 mm resulted in low drain outflow. Each plot was described by its own equation.

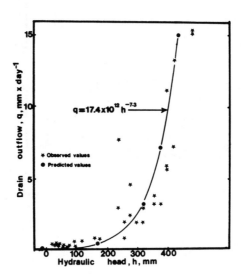

Fig. 4 Discharge vs Hydraulic Head Relationship using Observed and Predicted Values for Plot 3.

The drain spacing formula for unsteady state conditions and for a falling water table situation developed by Kraijenhoff Van De Leur (Dieleman 1974) was used to calculate the hydraulic conductivity of each plot. The equation can be written in the following form:

$$Kd_e = \frac{q}{h} \frac{L^2}{2\,\pi} \qquad\qquad (2)$$

where q = drain outflow
  h = hydraulic head
  L = drain spacing
  K = hydraulic conductivity
 de = equivalent depth.

The discharge-hydraulic head relationship is not constant; therefore K must be calculated for different q/h values. The slope of each curve, dq/dh, for each plot was determined from the theoretical curves. For different values of h, the hydraulic conductivity was then determined for each plot.

The mean hydraulic conductivity values of the plots as determined by three different methods is shown in Table 1. The arithmetic mean is shown with the standard deviation and C.V. Large C.V. values were to be expected because hydraulic conductivity values usually are log-normally distributed

(Dieleman and Trafford 1976; Beckett and Webster 1971). However, the distribution of the hydraulic conductivity values among plots for all hydraulic heads was found to be normal even with very high C.V. The median values were less than the arithmetic means and the geometric means were the least. This could have a significant effect when choosing a single value for hydraulic conductivity in designing a subsurface drainage system.

Table 1. Summary of the Mean Hydraulic Conductivity Values (m/d) for 10 Plots.

Hydraulic head mm	Arithmetic mean $\bar{X}$	Standard deviation S	Coefficient variability %	Median	Geometric mean
0	.111	.0832	75.0	.0936	.0820
100	.275	.207	75.2	.235	.205
200	1.034	1.030	99.6	.694	.648
300	3.971	4.572	115.1	2.454	2.182
400	21.517	29.923	139.1	11.097	9.343

Further statistical tests were performed to determine whether or not the lysimeter plots could be grouped and to establish differences among plots.

When using the Mann-Whitney U test to compare the hydraulic conductivity values of the first six plots with those of the remaining 4 plots (plots 9 and 12 were not used) significant differences were found between the two groups. Hydraulic conductivity value differences among all plots were found to be significant when using the Freidman two-way analysis of variance. Hydraulic conductivity values in plots 8, 10 and 11 were all significantly different from those in plots 1 and 4. These differences could justify the separation of the plots into two distinct groups or experimental blocks, namely plots 1 to 6 and 7 to 12.

Drainable Porosity

Drainable porosities were calculated from the ratio of drain outflow to water table drawdown for each plot. Only values obtained from the October hydrological event were considered because little or no evaporation occurred at that time. Unfortunately this limited the available values for discussion.

During the October event the water table level in most of the plots dropped from about 350 to 550 mm below ground surface. The drainable porosity values were determined for this layer. For most of the plots, it decreased with increasing depth. Lévesque and Millette (1977) report that for this organic soil the percentage fiber between .60 and 1.5 m from the soil surface is between 15 and 20% whereas near the surface they found 25 to 60%. The drainable porosity values ranged from 19 to 31% at 35 cm below the soil surface, and from 3.1 to 14% at about 550 mm below soil surface. It is probable that the drainable porosity is even less than 3% near the drain depth. Boelter (1974) found values ranging 12 to 57% for herbaceous peat and 8% for well decomposed peat.

These decreasing drainable porosity values with depth could explain the general low hydraulic conductivity near the drain depth. Another indication of low drainable porosity near the drain depth is visible from Fig. 4, which shows very little change in drain outflow for hydraulic heads of up to 200 mm. This is consistant throughout the plots. The decision on which drainable porosity value to use for subsurface drainage design again appears difficult. Moreover, as the soil is cultivated further degradation of the surface and compaction of the subsurface layers will decrease the drainable porosity values.

SUMMARY AND CONCLUSIONS

As part of an extensive drainage and water quality experiment twelve 6 x 15 m drainage lysimeter plots were constructed in an organic soil to determine the variability of soil water properties within a relatively small area. The data collected consisted of drain outflow from subsurface drains, water table readings and continuous precipitation at the site. With the use of these data and the Kraijenhoff Van De Leur equation, drainable porosity and hydraulic conductivity values were determined for each plot and compared. Differences in water table heights above drain among plots were not significantly different. However, monthly total drain outflow from each plot were compared and showed significant differences. Some plots yielded more water than others. Hydraulic conductivity values showed a decrease with depth in all plots and significant differences among plots. Using the arithmetic mean for estimating the hydraulic conductivity of all the plots may overestimate the actual value, the median and geometric mean being more conservative. Drainable porosity decreased with depth in all plots and showed large variations.

REFERENCES

Beckett, P.H.T. and R. Webster. 1971. Soil variability: a review. Soils & Fert. 34: 1-15.

Boelter, D.H. 1964. Water storage characteristics of several peats in situ. Soil Sci. Soc. Amer. Proc. 28: 433-435.

Boelter, D.H. 1965. Hydraulic conductivities of peats. Soil Sci. 100: 227-231.

Boelter, D.H. 1969. Physical properties of peats as related to degree of decomposition. Soil Sci. Soc. Amer. Proc. 33: 606-609.

Boelter, D.H. 1974. The hydrologic characteristics of undrained organic soils in the Lake States; in Histosols: their characteristics classification and use. SSSA Spec. Publ. No.6, 33-46.

Campbell, J.A., J.A. Millette and M. Roy. 1980. An inexpensive instrument for measuring soil water table levels. Can. J. Soil Sci. 60: 575-577.

Daniel, W.W. 1978. Applied nonparametric statistics. Houghton Mifflin Company, Boston, Mass. 503 p.

Dieleman, P.J. 1974. Deriving soil hydrological constants from field drainage tests. Drainage principles and applications, vol.III, 330-350. Inter. Inst. for Land Reclamation and Improvement, Wageningen, Netherlands.

Dieleman, P.J. and B.D. Trafford. 1976. Drainage testing. Irr. and Drainage, Paper 28, FAO, Rome, Italy.

Lévesque, M. et J.A. Millette. 1977. Description morphologique et aspects chimiques de la tourbière à laîches de Farnham, Québec. Naturaliste Can. 104: 511-526.

Nielsen, D.R., J.W. Biggar and K.T. Erb. 1973. Spatial variability of field measured soil water properties. Hilgardia 42: 215-259.

Warrick, A.W. and D.R. Nielsen. 1980. Spatial variability of soil physical properties in the field, 319-344, in Applications of Soil physics by Daniel Hillel. Academic Press, Inc. New York, N.Y. 385 p.

POTENTIAL FOR SUBSURFACE DRAINAGE IN

THE LOWER MISSISSIPPI VALLEY[1]

Cade E. Carter   Victor McDaniel   R. L. Bengtson
Member ASAE   Assoc. Member ASAE   Assoc. Member ASAE[2]

Subsurface drainage is recognized throughout the world as a practice that enhances crop production. It is used not only in humid areas but also in arid areas in conjunction with irrigation. The need for subsurface drainage is determined easily in some cases and very difficult to determine in others.

For example, if salt is a severe problem the potential benefits of subsurface drainage are recognized readily. In arid and semi-arid areas salt build up in the root zone may be a potential problem that, if not controlled by subsurface drains, could reduce crop yields for several years until the salt has been leached from the soil profile. Another case in which the potential for subsurface drainage may be recognized readily is that of a relatively short growing season like in the midwestern and northern sections of the U.S. Subsurface drainage is needed in those areas to provide good trafficability and to aid in warming the soil in the early spring so that seeds will germinate before being damaged by wet, cool conditions. In the cool humid areas the potential for subsurface drainage can be determined readily by comparing the value of high yielding crops planted in a timely manner to lower yielding crops that were planted too late.

In the humid south, however, soil salinity is not a serious problem because high rainfall keeps salt leached from the root zone. Getting into the field early in the spring for most crops is not a valid justification for subsurface drainage since the growing season is relatively long. The economic potential for subsurface drainage in the humid south is therefore a matter of whether subsurface drainage increases yields sufficiently to justify the costs of the drains. In some years, crop yields may not be increased by subsurface drainage due to droughts. Thus, the frequency of need also must be included in estimating the economic benefits.

The objectives of this experiment were to determine: a) the need for drainage, b) the response of crops to subsurface drainage, c) the estimated cost of subsurface drainage, and d) the economic potential for subsurface drainage in the lower Mississippi Valley.

PROCEDURE

The first step in this experiment was to determine the frequency of the water table rise into the root zone and how long it stayed. Observation wells to measure the water table were installed on Mhoon silty clay loam and Commerce silt loam in East Baton Rouge Parish, Louisiana, in 1974

[1]/Contribution from the U.S. Department of Agriculture, Agricultural Research Service, and the Louisiana Agricultural Experiment Station, Baton Rouge, La.

[2]/Agricultural Engineer, USDA-ARS; Instructor, LSU; and Assistant Professor, LSU; respectively, Baton Rouge, La.

(Figure 1). Measurements were made and recorded weekly through 1980. In 1977 another observation well was installed, also on Commerce silt loam, but in St. James Parish (Figure 1). This well was instrumented in 1978 with a water stage recorder that gave a continuous record of the water table. Also, in 1978 additional observation wells with recorders were installed on Baldwin silty clay in St. Mary Parish, Jeanerette silty clay loam in Iberia Parish, and Commerce silty clay loam in East Baton Rouge Parish (Figure 1). Water tables measured from each of these sites were plotted, and the number of days the water table was within 45 cm of the soil was determined.

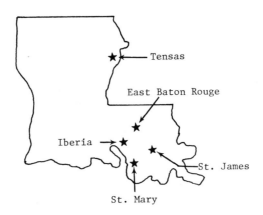

Fig. 1   Location of subsurface drainage
experiments in Louisiana
1974 - 1981

The second step in this experiment was to determine drain spacing needed to drain the soil. Drain spacings were estimated using the following from Schwab, et al, 1966:

$$S = \left[\frac{9Ktd_e}{f[\ln m_o(2d_e + m) - \ln m(2d_e + m_o)]} \right]^{1/2} \tag{1}$$

where:

S = drain spacing
K = hydraulic conductivity
d_e = equivalent depth
m = height of water table above the drain at midplane after time t
m_o = initial height of water table, m
t = time in days for water table to drop from m_o to m
f = drainable porosity

Hydraulic conductivities were determined using the auger hole method (Boersma, 1965) for each soil except Sharkey clay. Porosities for each site were measured in the laboratory from soil samples obtained from the field, with a few exceptions where published porosities were used (Vomocil, 1965, Lund and Loftin, 1960). Drainage systems were installed at the estimated spacing in most cases. Drains were also installed closer than was estimated by equation 1 at several sites and wider apart than was estimated by equation 1 at two sites.

The third step in this experiment was to determine the effectiveness of sub-surface drainage in lowering the water table and to determine crop response to the improved soil water regime that the drainage system provided. Sub-surface drainage systems were installed at several locations in Louisiana beginning in 1973 for these determinations. A description of each site is given in Table 1.

The fourth and final step in this experiment was to pool the information gathered and to estimate the economic potential for subsurface drainage in the lower Mississippi Valley. Factors used in estimating the economic potential for subsurface drainage varies from year to year. For example, the response of the crop to drainage varies due to climatic variations. The price growers receive for their crops varies depending upon supply and demand. Drain spacing required to drain the soil effectively varies from one soil to another due to many factors including soil texture, soil layering, etc. The cost of drainage varies depending upon drain spacing and cost of installation. Due to these variations, a nomograph was assembled that included a range of values for each factor so that a different combination of factors could easily be considered in estimating the economic potential of drainage. Yield response and estimated drain spacing from each experiment reported in this paper were used in the nomograph to estimate the economic potential for subsurface drainage.

RESULTS AND DISCUSSION

Water table data, collected to determine the frequency and duration of high water tables, showed a potential need for subsurface drainage since high water tables are known to reduce crop yields (Wesseling, et al, 1957, Williamson and Kriz, 1970, Tondreau, et al, 1976). In East Baton Rouge Parish in 1975, a wet year with 1720 mm of rainfall, the water table rose to within 45 cm of the soil surface of Commerce soil on 7 different occasions for a total of 179 days (Table 2). These data mean that the water table was within 45 cm of the soil surface almost 50% of the time in 1975. Although this example was specifically cited to illustrate the problem of high water tables for unusually long periods, data in Table 2 show that the water table was high 25% or more of the time many years. Only a few times, like in 1978 in St. James Parish, did the data indicate that drainage, perhaps, was not needed. A typical water table in an undrained area, from which data shown in Table 2 were obtained, is shown in Figure 2. The water tables at all sites fluctuated considerably, depending upon the frequency and amount of rainfall.

The effectiveness of subsurface drainage in lowering the water table was determined by comparing water levels like those shown in Figure 2 and Figure 3 where data are shown from areas without and with subsurface drainage, respectively. A further comparison was the frequency and duration of the water table in the root zone for areas with and without subsurface drains as shown in Table 3. An illustration of the effectiveness of subsurface drainage can be obtained from the 1979 St. James Parish data. Rainfall was above average that year resulting in unusually high water tables which sometimes rose above the soil surface. Subsurface drainage reduced the number of days the water table was within 45 cm of the soil surface by 48%. Similar comparisons for other locations can be made from the data in Table 3.

Crops responded favorably to subsurface drainage. In most cases, yields were higher from areas with low water tables (Table 4). An exception was in 1978 in St. James Parish where the area without subsurface drains yielded more than did the area with subsurface drains. This reversal in drainage-increases-yield trend was attributed to a drought during the summer. Yields from the different drain spacing treatments at the Tensas, Iberia, and St. Mary sites differed considerably. To illustrate this potential for increasing crop yields by subsurface drainage, yields from several drain spacings are reported in Table 4. Annual yields of sugar, soybeans, corn silage, and wheat were increased a maximum of 33, 39, 89, and 73 percent, respectively, by subsurface drainage.

Table 1. A description of experimental sites in Louisiana used in estimating the potential for subsurface drainage.

Site Location (Parish)	Soil Type	Drains Installed (year)	Drained Area (ha)	Drain Spacing (m)	Drain Depth (m)	Drain Slope (%)	Crop	Experimental Period (year)
Tensas	Sharkey c	1973	4.0	7.5 & 15	1.2	.15	Soybeans	1974-1977
St. James	Commerce sil	1976	5.0	24, 36, & 48	1.2	.17	Sugar	1977-1980
St. James	Commerce sil	1976	5.0	24, 36, & 48	1.2	.17	Wheat	1981
St. Mary	Baldwin sic	1977	1.6	14 & 28	0.9	.20	Sugar	1979-1980
Iberia	Jeanerette sicl	1977	4.2	14, 28, & 42	1.3	.20	Sugar	1980-1981
E. Baton Rouge	Commerce sicl	1975	2.0	10 & 20	0.9	.15	Corn Silage	1980-1981

Table 2. Frequency of Duration of the Water Table Within 45 cm of the Soil Surface for Undrained Areas.

| Location | Soil | Year | Water table within 45 cm of soil surface | | Annual Rainfall |
			Frequency (No)	Duration (Days)	(mm)
E. Baton Rouge	Commerce sil	1974	9	135	1517
E. Baton Rouge	Mhoon sicl	1974	8	81	1517
E. Baton Rouge	Commerce sil	1975	7	179	1720
E. Baton Rouge	Mhoon sicl	1975	7	113	1720
E. Baton Rouge	Commerce sil	1976	4	58	1246
E. Baton Rouge	Mhoon sicl	1976	4	60	1246
E. Baton Rouge	Commerce sil	1977	4	102	1960
E. Baton Rouge	Mhoon sicl	1977	6	153	1960
E. Baton Rouge	Commerce sil	1978	7	135	1488
E. Baton Rouge	Mhoon sicl	1978	5	96	1488
E. Baton Rouge	Commerce sil	1979	7	136	1795
E. Baton Rouge	Mhoon sicl	1979	8	135	1795
E. Baton Rouge	Commerce sil	1980	5	109	1686
E. Baton Rouge	Mhoon sicl	1980	8	102	1686
St. James	Commerce sil	1978	7	11	1612
St. James	Commerce sil	1979	21	92	1790
St. James	Commerce sil	1980	8	97	1892
St. James	Commerce sil	1981 (5 Mos)	2	10	439 (5 Mos)
St. Mary	Baldwin sic	1979	15	69	2022
St. Mary	Baldwin sic	1980	19	33	1940
Iberia	Jeanerette sicl	1980	6	47	1676
Iberia	Jeanerette sicl	1981	8	40	1145

Table 3. Water Table Response to Subsurface Drainage.

| Location (Parish) | Drain Spacing (m) | Year | Number of days water table was within 45 cm of soil surface | |
			Drained	Undrained
St. James	36	1978	5	11
St. James	36	1979	48	92
St. James	36	1980	17	97
St. James	36	1981	0	10
St. Mary	14	1979	30	69
St. Mary	14	1980	7	33
Iberia	28	1980	26	47
Iberia	28	1981	4	40
E. Baton Rouge	20	1981	6	63ᵃ

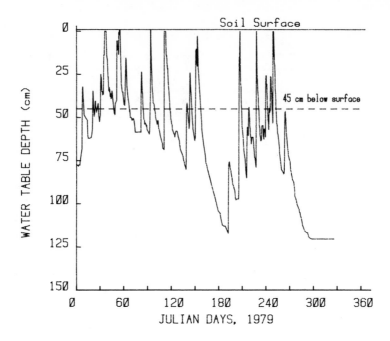

FIGURE 2. WATER TABLE FROM UNDRAINED COMMERCE SIL, ST. JAMES PARISH, LA.

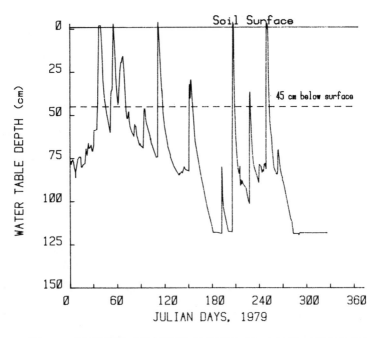

FIGURE 3. WATER TABLE FROM SUBSURFACE DRAINED (36m SPACING) COMMERCE SIL, ST. JAMES PARISH, LA.

Table 4. Crop Yields from Subsurface Drainage Experiments Conducted in
Several Louisiana Parishes.

Location (Parish)	Drain Spacing (m)	Year	Crop	Yields (kg/ha) Drained	Undrained
Tensas	7.5	1974	Soybean	2585	2347
Tensas	7.5	1975	Soybean	2174	1748
Tensas	7.5	1976	Soybean	1594	1143
Tensas	7.5	1977	Soybean	2422	2335
St. James	36.0	1977	Sugar	6244	5699
St. James	36.0	1978	Sugar	7493	8584
St. James	36.0	1979	Sugar	8179	6629
St. James	36.0	1980	Sugar	5189	5040
St. Mary	14.0	1979	Sugar	5800	4864
St. Mary	14.0	1980	Sugar	4455	3933
Iberia	28.0	1980	Sugar	5844	5165
Iberia	28.0	1981	Sugar	7973	6003
St. James	36.0	1981	Wheat	3475	2040
E. Baton Rouge	10.0	1980	Corn Silage	32647[a]	17971[a]
E. Baton Rouge	10.0	1981	Corn Silage	27249[a]	14887[a]

[a]60% moisture

Amortization was used to determine whether or not subsurface drainage was
feasible. If the benefits were sufficient to pay the costs of subsurface
drainage in less than 10 years (the actual life of a drainage system should
be much longer than 10 years) then the practice was considered feasible. To
make this determination, a nomograph that relates the variables involved in
this decision was used (Figure 4). The average annual crop yield increase
in kg/ha measured from each experiment was entered on the negative Y axis.
The point at which the yield increase crossed the associated crop value curve
provided the increased crop value in $/ha on the negative X axis. Next, the
drain spacing was entered on the positive X axis. The point at which it
crossed current drain cost/m line provided the drain installation cost in
$/ha on the positive Y axis. Finally the intersecting of the drain-installa-
tion costs and increased-crop-value provided the amortization. For example,
the average yield increase from the experiment with sugarcane in Iberia
Parish was 1523 kg/ha. If the price of sugar is $0.44 per kg, the increased
value of the crop is $670/ha. Drain spacing was estimated at 28 m; and if
the installation cost is $2.46/m, then installation cost is $879/ha. This
cost and the increased crop value intersects on the amortization curve at 1.3
years which is shown by dashed lines in Figure 4. A summary of the data from
each experiment is given in Table 5. The crop prices were from market
reports published in the newspaper during the summer of 1982.

The nomograph, Figure 4, is convenient for determining not only a close
estimate of the pay back period for a given set of conditions but also for
considering a range of conditions. For example, if world instead of domestic
price were used, would subsurface drainage be feasible with 28 m spacing and
a crop yield increase due to drainage of 1523 kg/ha annually (Iberia Parish
data)? This question can be answered easily using the same procedure as
before. With world price (crop value) of 16¢/kg the increased crop value is
$244, and amortization is 3.6 years.

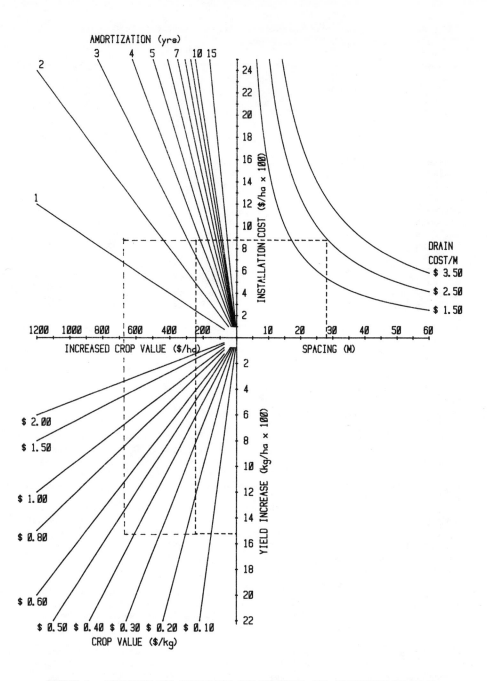

FIGURE 4.   NOMOGRAPH FOR ESTIMATING THE POTENTIAL FOR SUBSURFACE DRAINAGE.

Table 5. Summary of Data Used in Estimating the Potential for Subsurface Drainage.

Location (Parish)	Experiment conducted (year)	Crop	Average annual yield increase due to drainage (kg/ha)	Value of crop ($/kg)	Value of yield increase ($/ha)	Drain spacing (m)	Drain cost ($/ha)[a]	Amortization (year)
Tensas	1974–1977	Soybeans	297	.280	83	7.5	3280	39.5
Tensas	1974–1977	Soybeans	229	.280	64	15.0	1640	25.6
St. James	1977–1980	Sugar	429	.440	189	24.0	1025	5.4
St. James	1977–1980	Sugar	212	.440	93	36.0	683	7.3
St. James	1977–1980	Sugar	495	.440	218	48.0	512	2.4
St. Mary	1979–1980	Sugar	729	.440	321	14.0	1757	5.6
St. Mary	1979–1980	Sugar	179	.440	79	28.0	879	11.1
Iberia	1980–1981	Sugar	1509	.440	664	14.0	1757	2.6
Iberia	1980–1981	Sugar	1523	.440	670	28.0	879	1.3
Iberia	1980–1981	Sugar	1125	.440	495	42.0	586	1.2
E. Baton Rouge	1980–1981	Corn Silage	13529	.038	514	10.0	2460	4.8
E. Baton Rouge	1980–1981	Corn Silage	10802	.038	410	20.0	1230	3.0
St. James	1981	Wheat	1435	.140	201	( )[b]	683	3.4

[a] drain cost = $2.46/m

[b] 24, 36, and 48 m spacing

The data in Table 5 show that the 28 m drain spacing on Baldwin sic soil in St. Mary Parish would not be economically feasible for sugar production since the indicated amortization is more than 10 years. These data may be misleading since the low yield from this treatment was attributed to a serious Johnsongrass problem that developed. Grass control was not as effective in this treatment as in the other two treatments at this site.

It is interesting to note that the drain spacing treatment with the highest yield did not always have the quickest pay back period. For example, in the Iberia Parish experiment the 42 m spacing treatment's increase-in-yield-due-to-drainage was less than the 14 m spacing treatment, yet amortization for the 42 m spacing was less than half that of the 14 m spacing (Table 5).

Wheat was grown in 1981 in St. James Parish during a period when fields are normally fallow. Common practice in Louisiana is to grow three crops (two ratoons) of sugarcane then fallow the land after harvest in November until planting time the following August or September. In the St. James experiment, wheat was planted immediately after cane harvest in 1980 and was harvested in May of 1981 well in advance of sugarcane planting time. Wheat not only provided extra income; but it also provided cover, protecting the soil. Since only one year of data is available, it is not known whether yield increases like that in 1981 can be expected annually.

The amortization for drainage of clay soil for soybeans in Tensas Parish shows clearly that drainage is not economically feasible at today's drainage costs and soybean prices. The price of soybeans would have to be 77¢ per kg (almost 3 times the present price) in order for amortization to be less than 10 years for 15 m spacing.

With exception of Sharkey clay, the potential for subsurface drainage is very favorable based on criteria that drainage costs be amortized in less than 10 years. The analysis did not include, however, interest rates or returns on investment. The pay back periods indicate that subsurface drainage may be feasible at most locations even at present high interest rates.

If double cropping were practiced, the potential for subsurface drainage would be even greater than that indicated by the data in Table 5. The long growing season in the lower Mississippi Valley provides an excellent opportunity for doubling cropping, a practice that is gaining widespread attention. With crop responses to drainage like those reported here, subsurface drainage is expected to become a common practice in the lower Mississippi Valley for both single and double cropping schemes.

SUMMARY AND CONCLUSIONS

High water tables are a serious problem in the lower Mississippi Valley. These water tables can be lowered by subsurface drainage. Sugarcane, soybeans, wheat, and corn silage all responded favorably when the water tables were lowered with maximum annual yield increases of 33, 39, 89, and 73 percent, respectively. With only one exception, the value of the average annual crop yield increase due to drainage indicated that subsurface drainage systems could be amortized in a short period; thus a great economic potential for subsurface drainage exists in the lower Mississippi Valley.

REFERENCES

1. Boersma, L. 1965. Field measurement of hydraulic conductivity below a water table. Methods of Soil Analysis, Part 1. Amer. Soc. of Agron. Nomograph No. 9. Amer. Soc. of Agron., Pub. pp. 222-233.

2. Lund, Z.F. and L.L. Loftin. 1960. Physical characteristics of some representative Louisiana soils. Agr. Res. Serv., U.S. Dept. of Agr. 83 p.

3. Schwab, G.O., R.K. Frevert, T.W. Edminster and K.K. Barnes. 1966. Soil and Water Conservation Engineering. John Wiley & Sons, Inc., New York, N.Y. 683 p.

4. Tondreau, J.E., L.S. Willardson, B.D. Meek and L.B. Grass. 1976. Soil aeration response to a fluctuated water table. Third National Drainage Symposium. pp. 130-134.

5. Vomocil, J.A. 1965. Porosity. Methods of Soil Analysis, Part 1. Amer. Soc. of Agron. Nomograph No. 9. Amer. Soc. of Agron., Pub. pp. 299-314.

6. Wesseling, J., W.R. van Wijk, M. Fireman, B.D. van't Woudt and R.M. Hagan. 1957. Land drainage in relation to soils and crops. Drainage of Agr. Lands. Amer. Soc. of Agron., Pub. 7: 461-611.

7. Williamson, R.E. and G.J. Kriz. 1970. Response of agricultural crops to flooding, depth to water table and soil gaseous composition. TRANS. of the ASAE 13(2): 216-220.

SHALLOW-WATER-TABLE CONCEPT FOR DRAINAGE DESIGN IN SEMIARID AND SUBHUMID REGIONS

E. J. Doering, L. C. Benz, and G. A. Reichman
Member   Member
ASAE   ASAE

Maximum plant growth requires, among other factors, a particular soil-water regime which may exist naturally or be provided by appropriate irrigation and drainage. Irrigation and drainage technologies must not be treated as separate entities however, because they interact with each other in the soil-water-plant system.

Drainage technology has developed around 2 basic needs: (1) to ensure aeration and trafficability for agricultural soils, and (2) to provide for salinity control. From an engineering standpoint, drainage is usually viewed in terms of physical processes involved with the flow of water in porous media, indigenous physical constraints, and economies of construction. As a consequence, the goal usually has been to design systems that provide as much drainage as possible, i.e., as deep a water table as possible with a minimum of capital investment.

Fine-textured agricultural soils in humid regions are often too wet because they have slowly permeable layers at shallow depths. Design depths for drains in these soils are restricted. Sandy soils in humid regions can have reduced productivity from either being too wet or too dry because of extremes and irregularities of precipitation events over time, and they can be subjected to both of those soil-water extremes in the same crop season. Unless a restrictive layer exists at a shallow depth beneath sandy soils, the prevailing drainage practice has been to install deep drains to minimize the length of drain per hectare. The result is that excessive drainage has often been provided and crop production is reduced because of water stress when rainfall events do not occur regularly during the growing season.

Drains for irrigated lands in arid regions are commonly installed as deep as is physically and economically feasible, i.e., to depths of 3 meters or more in southwestern United States. Ochs et al. (1980) stated, "An ideal condition in an irrigated area would be one where the water table is deep and the soil is so constituted that there is unrestricted downward movement of excess water and salt from the root zone". This deep drainage concept may be justified for arid regions because saline water tables usually develop and there is danger that evapotranspiration (ET) will cause salts to accumulate in the root zone.

Humid, subhumid, and semiarid regions need irrigation only to supplement rainfall. Efficient use of rainfall requires that water storage capacity be available when rainfall events occur. A shallow nonsaline water table is particularly beneficial for supplementally irrigated lands because plants can extract some water from that water table, which adds flexibility to irrigation schedules and makes possible the more efficient

The authors are: E. J. DOERING, Agricultural Engineer, L. C. BENZ, Agricultural Engineer, and G. A. REICHMAN, Soil Scientist, USDA-ARS, Northern Great Plains Research Center, P. O. Box 459, Mandan, ND.

use of natural precipitation. It is the purpose of this paper (1) to show
how application of the shallow water-table concept to coarse-textured
irrigated soils significantly reduces irrigation requirements for corn and
sugarbeets, (2) to use the Donnan steady state equation to show how much
drain spacing is reduced by changing from a deep to a shallower drain
system, and (3) to use published drain construction cost data to compare
the initial costs for those 2 drain systems.

DRAINAGE REQUIREMENTS FOR CROPS

Various crops respond differently to specific soil-water environments; so
the key element of the shallow water-table concept is the drainage
requirement of the crop being grown. The term "drainage requirement" is
often used to specify the depth of water removed per day. In this paper,
however, "crop drainage requirement" refers to the degree of drainage
needed to provide optimum growing conditions. As such, it defines the
minimum water-table depth which satisfies plant requirements of rooting
volume, nutrition, soil temperature, and aeration along with such
considerations as irrigation management, water resource conservation,
trafficability, soil salinity, and environmental quality. Doering et al.
(1977) showed that the drainage requirement for corn grown on sandy soils
in North Dakota was about 1 m.

Because of natural variability of agricultural soils, climatic
uncertainties, and a scarcity of precise drainage requirement information,
Hiler (1969) and Ochs et al. (1980) concluded that drainage engineers
could not design the optimum drainage system to maximize agricultural
returns. However, controlled water-table depths have been used in
Minnesota, Indiana, Ohio, and South Carolina to provide subirrigation for
crops (Criddle and Kalisvaart 1967; Doty and Parsons 1979). Beneficial
effects of the water table have been demonstrated by several experiments
that have been summarized by Williamson and Kriz (1970). Reichman et al.
(1977) showed that yield and quality of sugarbeets were both high for the
shallowest time-averaged water-table depth (1.5 m) available for their
experiments on sandy soils in North Dakota.

MATERIALS AND METHODS

The research reported herein was conducted in both field plots and large
nonweighing lysimeters. The field-plot research was conducted on sandy
soils of the Hecla-Arveson-Fossum association. Hecla soils are Aquic
Haploborolls, sandy mixed; Arveson soils are Typic Calciaquolls,
coarse-loamy, frigid; and Fossum soils are Typic Haplaquolls, sandy, mixed
(calcareous), frigid; (USDA-SCS 1981). Before initiation of the
experiments, corn was grown on the entire 32-ha experimental area and plot
locations were selected on the basis of plant growth evaluations. The
natural topography of the area was used in combination with the cone of
depression around a pumped well to provide 3 water-table depth
treatments. Water-table depth treatments were established as main blocks
with 4 irrigation treatments randomized in each of 4 replicates and with
the crops imposed as split plots. Corn and sugarbeets were grown on the
same plots for 3 years. The 3-year, time-averaged (May 15 through
September 30) water-table depths were 1.5 m, 2 m, and 2.3 m and were
designated shallow, medium, and deep, respectively. The water-table depth
increased with time during the growing season each year.

The amounts of irrigation water (IW) required to provide the irrigation
treatments were determined weekly by:

$$IW = C (ET_p \times K_c) - P + D \qquad [1]$$

where C is the irrigation level, ET_p is potential evapotranspiration calculated by the Jensen-Haise equation (Burman et al. 1980) adapted for North Dakota conditions (Follett et al. 1973), K_c is a crop coefficient that accounts for stage of crop development (Burman et al. 1980), P is measured effective precipitation, and D is a calculated soil-water deficit carried over from the previous week. The use of C equal to 0.5, 1.0, and 1.5 provided 3 separate irrigation treatments based on calculated water requirements for corn. A nonirrigated treatment was also included. Daily precipitation events less than 2.5 mm were considered noneffective and were ignored. Calculated IW amounts less than 7.6 mm were not applied and were carried forward as D for the following week. If large precipitation events caused IW to be negative, that negative amount was ignored because the excess water was assumed to drain through these sandy profiles and to have no effect on the following irrigation period. Irrigation was applied with a rotating-boom plot irrigator (Bond et al. 1970).

The first year, N as ammonium nitrate was broadcast at a rate of 224 kg/ha for corn and 112 kg/ha for sugarbeets. Sufficient N fertilizer was applied the second and third years to bring the soil N (NH_4-N plus NO_3-N) to an available N level of 224 kg/ha for corn and 112 kg/ha for sugarbeets. Concentrated superphosphate and potassium sulfate were broadcast the first year at rates of 56 kg P/ha and 111 kg K/ha. The second and third years, P was banded 5 cm below and 5 cm to the side of each row at a rate of 33 kg/ha. Zinc sulfate was banded beside and below each row at a rate of 5.6 kg Zn/ha the first year and 4.4 kg Zn/ha the following years. Corn grain yields are reported on a dry matter basis. Sucrose percentage of the sugarbeet roots was determined by a polariscopic method similar to that outlined by the Association of Official Agricultural Chemists (1955).

Twelve automatic, nonweighing lysimeters were constructed (Reichman et al. 1979), each with a surface area of 11.9 m^2 (2.44 m by 4.88 m). Four constant water-table depths (46, 101, 155, and 210 cm) and 3 irrigation levels (C = 0.3, 0.8, and 1.3 for equation 1) were used in a factorial experiment. Irrigation amounts were calculated and applied as needed each Monday and Friday instead of once each week as for the field plots. Corn and sugarbeets were grown the first and second years, respectively, with each of the 4 rows being harvested separately and treated as replicates.

The Donnan steady state equation (USDI-BR 1978) was used to calculate drain spacings for shallow (1.5 m) and deep (2.4 m) drains with 1.0- and 1.2-m midpoint water-table depths, respectively.

RESULTS AND DISCUSSION

It is hypothesized that yield is related to water-table depth for a particular irrigation level as shown in Fig. 1. Thus, the water-table depth that produces the maximum yield describes the drainage requirement for that crop. Corn grain and sugarbeet sucrose yields (replicate means) for the field plots are presented as functions of water-table depth with irrigation levels as parameters in Figs. 2 and 3, respectively. The yield lines tend to slope upward to the left between the shallow and medium water-table depths, indicating that the water table for these field experiments was too deep to define the crop drainage requirements. Hence, the lysimeter experiments with shallow water tables were required.

Corn grain and sugarbeet sucrose yields (replicate means) for the lysimeters are shown as functions of water-table depth using irrigation levels as parameters in Figs. 4 and 5, respectively. Yields of both crops were significantly reduced for the 46-cm water table, but irrigation had little or no effect on production for the 46-cm water table. The smooth

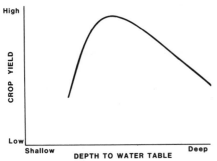

FIG. 1. *Theoretical relationship for crop yield and water-table depth.*

curves shown in Figs. 4 and 5 were generated by fitting a cubic equation to the data. The maximum crop yield values for all 3 curves are approximately equal, but the maximum for the 1.3 irrigation level occurs for a deeper water table than it does for the 0.3 and 0.8 irrigation level. These results show that the drainage requirement for corn and sugarbeets (as defined by Fig. 1) is about 1 m, and that drainage requirement depends on irrigation management as well as on water-table depth, i.e., that irrigation and drainage interact as parts of the total water management system. We can not explain why the corn yield with 0.3 irrigation was higher for the 2-m water-table depth than for the 1.5-m water-table depth.

By setting the derivative of each cubic equation equal to zero, the water-table depth at which the maximum yield occurred was obtained. We

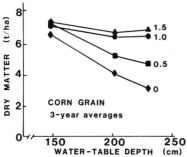

FIG. 2. *Corn yield as a function of water-table depth for 3 irrigation levels (0.5, 1.0 & 1.5) and no irrigation (0) in field plots.*

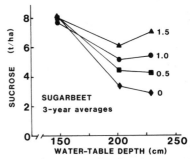

FIG. 3. *Sucrose as a function of water-table depth for 3 irrigation levels (0.5, 1.0 & 1.5) and no irrigation (0) in field plots.*

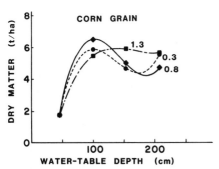

FIG. 4. *Corn yield as a function of water-table depth for 3 irrigation levels in lysimeters.*

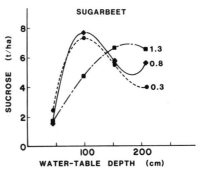

FIG. 5. *Sucrose yield as a function of water-table depth for 3 irrigation levels in lysimeters.*

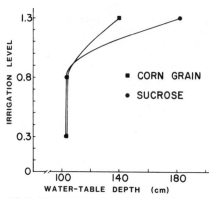

FIG. 6. Interaction of water-table depth and irrigation level for maximum production.

acknowledge that these crop drainage requirement values are not exact values and that probable confidence intervals coannot be inferred from these experiments. In spite of this limitation, the interaction of irrigation and drainage is shown by a graph in which the irrigation level and water-table depth values corresponding to maximum yields were used as coordinates (Fig. 6). As water-table depth increases beyond 100 cm, the irrigation requirement increases significantly. In these lysimeter experiments, irrigation level 0.8 required about 20 cm more water than did level 0.3, and level 1.3 required about 20 cm more water than did level 0.8. With a water table at the 100-cm depth, the 0.3 irrigation level would have the lowest annual irrigation cost. Placing drains 2.4 m deep in these soils would result in a probable water-table depth of about 2 m early in each growing season -- requiring between 30 and 40 cm of additional irrigation per season, compared to the 1 m water table, to obtain comparable maximum yields. This additional irrigation wastes both water and the energy to apply it (Benz et al. 1981).

Reductions in the irrigation requirement for sandy soils reduces the rigidity of irrigation schedules. Less frequent irrigation reserves more storage capacity in the upper root zone for precipitation which increases the potential for efficient use of precipitation. A shallow water table eliminates or significantly reduces the potential for deep percolation, which increases overall water-use efficiency (Benz et al. 1978).

In a companion (not yet published) field-plot study, we found that net gains of N from the mineralization of organic matter and recovery of applied N by the crop were both greatest over the shallow water table. Increased irrigation required for the medium and deep water-table depths resulted in decreased N recovery by the crops and in lower N-use efficiencies. Thus, the shallow water table enhances N-use efficiency.

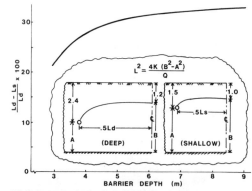

FIG. 7. Percent reduction in drain spacing when drain depth & midpoint water-table depth are changed from deep to shallow as shown in the inset.

Shallower placement of drains requires reduced drain spacings. The percent reduction in spacing when drain depth is decreased from 2.4 to 1.5 m is shown as a function of barrier depth in Fig. 7. For these calculations, the water-table depths midway between drains were 1.2 and 1.0 m, respectively. The geometries of the two drain systems are given in the inset of Fig. 7. The Donnan steady state equation was used to calculate drain spacing (L), and the same values for hydraulic conductivity (K) and drainage coefficient (Q) were used for each comparison. Different values for K and Q give different drain spacings, but the percent difference between drain spacing for particular drain depth and midpoint water-table depth combinations is a function of barrier depth only.

The two drain systems described in Fig. 7 may appear to be too similar to be called "deep" and "shallow". However, agricultural drains continue to flow during much of winter in the northern half of the United States and in southern Canada. By early spring, the water table will have been drawn down to about drain depth. Even with a water-table rise of 50 to 60 cm because of snowmelt and spring rains, the deep drainage system (2.4 m) will have a water-table depth of about 1.8 m at planting time and irrigation requirements will be high for much of the growing season. A similar water-table rise for the shallow drainage system (1.5 m) will bring the water table to about the desired 1-m depth for maximum crop production with minimum irrigation.

The following analysis is based on drainage geometries shown in the inset of Fig. 7 and the Donnan steady state equation. Changing from the deep to the shallow drainage system results in a drain spacing reduction of 30% for a barrier depth of 5 m. A 30% spacing reduction results in a 43% increase in length of drain per unit area of land for the shallow drains compared to the deep drains. Installation costs are not independent of drain depth because machine size, power requirements, and drain diameter all increase as drain depth increases. Christopher and Winger (1975) showed that construction costs for comparable gravel-enveloped drains were $11.55 and $8.33 per m for installation depths of 2.4 and 1.5 m, respectively. Based on those data and the drainage geometries shown in Fig. 7, the per ha construction cost for shallow (1.5 m) and deep (2.4 m) drains would be equal for a barrier depth of about 4 m, and shallow drains would be less expensive than deep drains for barrier depths less than 4 m. For a barrier depth of 9 m, the per ha construction cost for deep drains (2.4 m) would be only about 8% less than for shallow drains (1.5 m). For this example, construction cost savings for deep drains, as compared to shallow drains, would be more than offset by the increased annual operating costs associated with the increased irrigation requirements for crops grown on sandy soils in nonarid regions. Applicability of similar criteria to other soils is not known because drainage requirements for crops grown on medium- and fine-textured soils have not been determined.

Shallow drainage requires efficient irrigation management to ensure that waterlogging of the root zone does not occur. Deep drainage provides a buffer against irrigation inefficiencies of all kinds. We believe (1) that future energy costs and competition for irrigation water will require that drainage systems be designed to minimize irrigation requirements, and (2) that shallow water tables will become important water resources for growing crops.

CONCLUSIONS

Water management experiments in field plots and lysimeters have shown that the drainage requirement is about 1 meter for corn and sugarbeets grown on sandy soils in North Dakota and that irrigation requirements for comparable yields increased significantly as the water-table depth increased. An increase in the irrigation requirements without corresponding increases in crop production translates directly to higher production costs and inefficient use of water and energy. Thus, deep drainage correlates to over-drainage for sandy soils in semiarid, subhumid, and humid regions.

Calculations using the Donnan (steady state) equation for drain spacing and 2 combinations of drain depth (2.4 and 1.5 m) and midpoint water-table depth (1.2 and 1.0 m) show that the drain spacing ratio for particular drain geometries is a function of barrier depth only. Based on USDI-BR data for drain construction cost, the cost per ha for shallow (1.5 m) and deep (2.4 m) drains described herein would be equal when the barrier depth was about 4 m. For shallower barrier depths, the shallow drains would cost less to construct than the deep drains. One-time construction cost savings that may result because deep drains can be spaced farther apart than shallow drains may be more than offset by recurring annual losses because of reduced yields or increased operating costs.

REFERENCES

1. Association of Official Agricultural Chemists. 1955. Methods of Analysis, 8th Edition. Association of Official Agricultural Chemists, Washington, DC.

2. Benz, L. C., G. A. Reichman, E. J. Doering and R. F. Follett. 1978. Water-table depth and irrigation effects on applied water-use-efficiences of three crops. TRANS. of the ASAE. 21(4):723-728.

3. Benz, L. C., E. J. Doering and G. A. Reichman. 1981. Watertable management saves water and energy. TRANS. of the ASAE. 24(4):995-1001.

4. Bond, J. J., J. F. Power and H. M. Olson. 1970. Rotating-boom plot irrigator with offset mounting. TRANS. of the ASAE. 13(1):143-144, 147.

5. Burman, R. D., P. R. Nixon, J. L. Wright and W. O. Pruitt. 1980. Water requirements. Chapter 6 In Design and Operation of Farm Irrigation Systems. M. E. Jensen (ed.). ASAE Monograph Number 3:189-232.

6. Christopher, J. N. and R. J. Winger, Jr. 1975. Economical drain depth for irrigated areas. Paper presented at 1975 Irrigation and Drainage Specialty Conf. of ASCE, Logan, UT.

7. Criddle, W. D. and C. Kalisvaart. 1967. Subirrigation systems. In Irrigation of agricultural lands. R. M. Hagan, H. R. Haise and T. W. Edminster (eds.). Am. Soc. of Agron., Madison, WS. Agronomy Number 11:905-921.

8. Doering, E. J., G. A. Reichman, L. C. Benz and R. F. Follett. 1977. Drainage requirement for corn grown on sandy soil. Proc. of 3rd Natl. Drain. Symp., "Drainage for Increased Crop Production and a Quality Environment." ASAE Publication 1-77:144-147.

9. Doty, C. W. and J. E. Parsons. 1979. Water requirement and water-table variations for a controlled and reversible drainage system. TRANS. of the ASAE. 22(3):532-536, 539.

10. Follett, R. F., G. A. Reichman, E. J. Doering and L. C. Benz. 1973. A nomograph for estimating evapotranspiration. J. Soil & Water Conserv. 28(2):90-92.

11. Hiler, E. A.   1969.   Quantitative evaluation of crop-drainage
    requirements. TRANS. of the ASAE. 12(4):499-505.

12. Ochs, W. J., L. S. Willardson, C. R. Camp, Jr., W. W. Donnan, R. J.
    Winger and W. R. Johnston. 1980. Drainage requirements and systems.
    Chapter 7 In Design and Operation of Farm Irrigation Systems. M. E.
    Jensen (ed.). ASAE Monograph Number 3:235-277.

13. Reichman, G. A., E. J. Doering, L. C. Benz and R. F. Follett. 1977.
    Effects of water-table depth and irrigation on sugarbeet yield and
    quality. J. Am. Soc. Sugar Beet Tech. 19:275-287.

14. Reichman, G. A., E. J. Doering, L. C. Benz and R. F. Follett. 1979.
    Construction and performance of large automatic (non-weighing)
    lysimeters. TRANS. of the ASAE. 22(6):1343-1346, 1352.

15. U. S. Department of Agriculture, Soil Conservation Service. 1981.
    Soil Series of the United States, Puerto Rico, and the Virgin Islands:
    Their Taxonomic Classification. U. S. Government Printing Office,
    Washington, DC.

16. U. S. Department of Interior, Bureau of Reclamation. 1978. Drainage
    Manual. U. S. Government Printing Office, Washington, DC. 286 pp.

17. Williamson, R. E. and George J. Kriz. 1970. Response of agricultural
    crops to flooding, depth of water table and soil gaseous composition.
    TRANS. of the ASAE. 13(2):216-220.

DRAINAGE: AN INCREMENTAL NET BENEFIT APPROACH FOR OPTIMAL DESIGN

D. S. Durnford   T. H. Podmore   E. V. Richardson
Associate Member ASAE   Member ASAE

Optimal design of a drainage system involves two very distinct considerations. The first of these is the physical consequences of drain installation and the prediction of the impacts of artificial drainage on the soil-plant-water interrelationships. The second aspect of determining an optimum drainage system is economic. This area has received a great deal less attention in the literature than the first aspect. However, in view of the enormous investment being made in agricultural drainage systems throughout the world, new drainage systems should be close to being economically optimum and, certainly, they should be economically feasible.

The methodology presented in this paper is an attempt to provide a workable procedure which can be used to identify economically optimum subsurface drainage systems in irrigated areas. An optimal drain system is defined as one which maximizes the difference between the value of increased crop yields attributable to drain installation and the cost of the drains. The method presented is unique, to the authors' knowledge, in that it combines a simulation model with an optimization routine which minimizes a cost function. This eliminates the need to evaluate a large number of drain spacing and depth combinations and also makes determination of an optimal system more traceable.

The procedure could be adapted to several general situations but was developed using open and closed relief drainage systems proposed for arid, irrigated areas. Closed drains will be emphasized in this paper. It is assumed that the topography of the area is relatively flat and high water tables are a general condition of the area caused by high applications of irrigation water and insufficient natural drainage. The purpose of the drains, then, is to limit the net buildup of the water table during the crop season, allowing more effective leaching, reducing the frequency and extent of nonaerated conditions in the root zone and reducing the upflow of salts to the root zone if the groundwater is saline.

GENERAL PROCEDURE

Figure 1 outlines the general procedure used. In the main loop of the computer model, the variable HMIN is incremented. HMIN is the minimum midspan depth to water from ground surface during the crop season and represents the system level or level of protection afforded by the drain system. For a given irrigation schedule, soil properties and field geometry, the frequency and extent of nonaerated conditions in the root zone decrease, upflow of salts from a more saline groundwater table decreases and the effectiveness of leaching increases as HMIN increases. Therefore, HMIN is directly related to the performance level of the drain system.

The authors are: D. S. DURNFORD, Assistant Professor, Dept. of Civil Engineering, T. H. PODMORE, Associate Professor, Dept. of Agricultural and Chemical Engineering, and E. V. RICHARDSON, Professor, Dept. of Civil Engineering, Colorado State University.

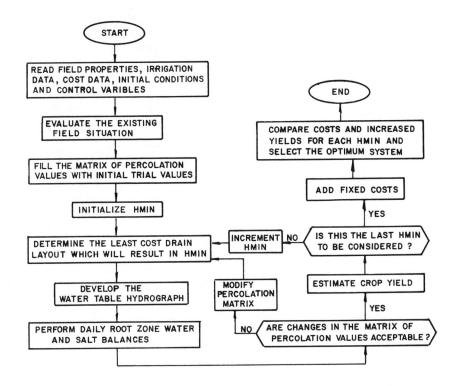

Fig. 1 General Procedure

At each system level defined by HMIN, there are an infinite number of drain spacing and depth combinations which will result in the same value of HMIN. However, there exists a unique least cost combination. This least cost combination of drain spacing and depth is identified by minimizing a cost function subject to constraints which approximate the trade-off between the drain spacing and depth required to result in a minimum depth to water of HMIN. Because the drain spacing and depth required are functions of the quantity of water percolating through the soil profile to the water table, which is, in turn, a function of the drain spacing and depth chosen, an iteration process is used to determine appropriate percolation values.

Water and salt balances of the root zone are performed on a daily basis. Upflow of water and salts from the water table, percolation to the water table and evapotranspiration are incorporated. In addition, root growth is retarded if high water table conditions exist and effective root depths are truncated if nonaerated conditions in the root zone persist for a specified number of days (depending on the crop's tolerance of nonaerated conditions). From these daily root zone balances, the effects of drain installation on the plant's environment and, hence, crop yield can be approximated.

For each system level, a minimum cost drain system and a corresponding crop yield are determined. Assuming increased crop yields represent the benefits derived, costs and benefits can be plotted against HMIN, the system level, and an optimum identified. Figure 2 illustrates an example of the type of output that would be expected.

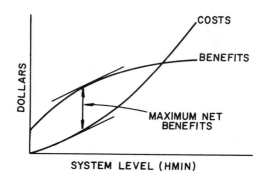

Fig. 2   Example Curve Showing Relationship between
System Level, Benefits and Costs

COST MINIMIZATION

The least cost drain depth and spacing combination which will result in the
required minimum midspan depth to water is determined at each level of HMIN
by a simple optimization routine. Generally, in determining drain layouts, a
drain spacing is calculated using an appropriate drain design equation so that
the design requirement (HMIN in this case) is satisfied. By minimizing a cost
function with the drain design equation being the major constraint, both the
drain depth and drain spacing are chosen. This drain depth and drain spacing
combination is the least cost combination which will satisfy the design
requirement.

The cost function, Z, assumed in the case presented here for closed drains,
in terms of dollars per unit area per year, is:

$$Z = \frac{i(C_1 + C_2\phi + C_3\phi^2)}{L} + \frac{iC_4}{L} + \frac{C_5}{L} + \frac{iC_6}{DPM*L} + \frac{iC_7}{L\ell} + \frac{i(C_8 + C_9 d)}{L} + \frac{iC_{10}}{L\ell} \tag{1}$$

where

d = depth of the center line of the drain
below ground surface

L = drain spacing, center to center

i = a rate which converts a present cost to an
annual cost for the expected life of the
drain

ϕ = drain diameter

ℓ = length of the drains

$C_1 + C_2\phi + C_3\phi^2$ = cost of the drain expressed as a quadratic
function of the diameter per unit length

C_4 = cost of the filter per unit length of drain

C_5 = cost of maintenance per unit length of
drain per year

C_6 = cost of one manhole

DPM = length of drain per manhole

C_7 = fixed cost of installing the drains, e.g., the cost of equipment use

$C_8 + C_9 d$ = cost of installation per unit length of drain expressed as a linear function of depth. The installation cost is assumed a function of the quantity of soil excavated.

C_{10} = cost of the drain outlet structure per outlet

The problem is constrained by restrictions or relationships between the variables. By introducing non-negative slack variables, ψ_i, i=1,5, in the constraints, the problem can be formulated in general form as:

$$\text{Min } \{Z = Z(L, d, \phi)\}$$

subject to

$$f_1 = d(L) = 0 \qquad\qquad f_5 = d_{max} - d - \psi_3 = 0$$

$$f_2 = \phi(d, L) = 0 \qquad\qquad f_6 = d - \psi_4 = 0$$

$$f_3 = L_{max} - L - \psi_1 = 0 \qquad\qquad f_7 = \phi - \phi_{min} - \psi_5 = 0$$

$$f_4 = L - \psi_2 = 0 \qquad\qquad \psi_1, \psi_2, \psi_3, \psi_4, \psi_5, \geq 0 \qquad (2)$$

The major constraint, f_1, is the required relationship between the drain spacing and the drain depth. In this particular case, it is assumed that the drain spacing and depth combination is adequately defined by the dynamic equilibrium equation (McWhorter 1977) and a specified value of HMIN. In the computer program, an approximation of the dynamic equilibrium equation is used in the constraints and checked elsewhere in the program. The drain diameter, ϕ, is eliminated entirely from the objective function and constraints by replacing it with an expression derived in terms of the other design variables. The minimization problem also incorporates convergence at the drain using criteria developed by Moody (1966) and the definition of the effective flow depth is included as a constraint. The minimization problem in the program is formulated as:

$$\underset{L, d, D}{\text{Min}} \quad \left(Z = \frac{C_1^*}{L} + \frac{C_2^* D^{3/8} (d - HMIN)^{3/8}}{L^{11/8}} + \frac{C_3^* D^{3/4} (d - HMIN)^{3/4}}{L^{14/8}} + \frac{C_4^* d}{L} \right)$$

subject to:

$$f_1 = d - HMIN - \frac{8}{\pi^2 K_1 Sya} \sum_{i=1}^{m} I_i \frac{\exp\frac{-\pi^2 \alpha \tau_i'}{L^2}}{1 - \exp\frac{-\pi^2 \alpha T}{L^2}} - \frac{I_m}{Sya}\left(1 - \frac{1}{K_1}\right) = 0$$

and

$$f_2 = D - d_e - \sum_{i=1}^{m} \frac{I_i L^2}{12 KDT} = 0 \qquad (3)$$

where

$$D = \text{effective flow depth}$$

$$d_e = \text{equivalent depth}$$

$$C_1^*, C_2^*, C_3^* \text{ and } C_4^* = \text{combined forms of the cost coefficients}$$

$$K_1 = \text{a shape factor}$$

$$K = \text{hydraulic conductivity}$$

$$Sya = \text{apparent specific yield}$$

$$m = \text{number of irrigation events}$$

$$I_i = \text{depth of water added to the water table from event } i$$

$$\alpha = \text{aquifer response factor} = KD/Sya$$

This form of the minimization problem has one degree of difficulty although the constraints are nonlinear. To circumvent the difficulty imposed by the nonlinearity of the constraints, the objective function and two constraint equations are differentiated with respect to the variables L, D, and d and evaluated at a point, say X_0. The point X_0 is a consistent, feasible point in the domain. If L is taken as the decision variable and D and d, the state variables, the following 3 x 3 system of linear, simultaneous equations can be solved for the gradient dZ/dL at X_0:

$$
\begin{bmatrix}
(\frac{\partial Z}{\partial L})_{X_0} & (\frac{\partial Z}{\partial d})_{X_0} & (\frac{\partial Z}{\partial D})_{X_0} \\
(\frac{\partial f_1}{\partial L})_{X_0} & (\frac{\partial f_1}{\partial d})_{X_0} & (\frac{\partial f_1}{\partial D})_{X_0} \\
(\frac{\partial f_2}{\partial L})_{X_0} & (\frac{\partial f_2}{\partial d})_{X_0} & (\frac{\partial f_2}{\partial D})_{X_0}
\end{bmatrix}
\begin{Bmatrix}
1 \\
\frac{dd}{dL} \\
\frac{dD}{dL}
\end{Bmatrix}
=
\begin{Bmatrix}
\frac{dZ}{dL} \\
0 \\
0
\end{Bmatrix}
\tag{4}
$$

If the gradient, dZ/dL, is found to be positive, the spacing is decreased, and conversely, if dZ/dL is negative, the spacing is increased. Selection of new trial spacings, evaluation of consistent values of D and d and solution of the matrix equation is repeated until a minimum value of the cost function is found.

A similar minimization problem has been developed for open ditch drains. Drain cost is assumed to be the sum of the excavation costs, maintenance costs and the value of the land taken out of production. The design variables are the drain depth and spacing, the effective flow depth of the groundwater, the bottom width of the ditch and the flow depth in the ditch.

BENEFIT ANALYSIS

Benefits derivable from installation of a drain system are difficult to quantify. Hiler (1977) and Sieben (Bouwer 1974) have proposed methodologies to quantify the effects of a fluctuating water table on crop yields. However, both are more applicable to humid rather than arid regions. In arid, irrigated agriculture, salinity is an almost universal problem and excess water is applied for leaching. If the water table is allowed to rise too high, upflow of water and salts from a saline groundwater table, ineffectiveness of leaching, nonaerated conditions in the root zone, and truncation of the root system may occur. Although any method which quantifies crop yield as a function of the drain system could be used in the general procedure discussed previously, an attempt has been made to quantify crop yields in terms of those factors which are generally the impetus for drain installation in arid regions. The methodology used in the program would not be expected to be adequate for a true crop yield evaluation but it does provide a means to quantify the effects of root zone soil moisture levels (both soil moisture deficits and soil moisture excesses), root zone salinity levels, and effective root depths on crop yield.

The yield of a particular crop is assumed to be directly related to plant growth which is assumed to be a direct function of its aerial and root environment. Further, the aerial environment and all aspects of the root environment except the soil moisture status, salinity status and location of the water table are assumed uncontrollable or nonlimiting to growth. In general algebraic form, yield is then a function of an index, S, with all other factors, K, assumed constant, i.e.,

$$Y = Y \; [S/K] \tag{5}$$

The index, S, is defined as the total integrated soil moisture suction in the root zone (Wadleigh and Ayers 1945; Bresler and Yaron 1972) and is an explicit function of the water content in the soil, the salinity of the root zone and the effective root depth. Assuming the plant responds to average values integrated over time and space but this response varies with the crop's growth stage, S is defined for each growth stage, i, as

$$S_i = \frac{1}{T_i} \int_0^{T_i} \frac{1}{Z(t)} \int_0^Z s(z,t)dzdt \tag{6}$$

where

$s(z,t) = \tau(\theta) + \pi(c)$; the sum of the matric suction,

$\tau(\theta)$, and the solute suction, $\pi(c)$, in bars

$Z(t)$ = the effective root depth

T_i = the length of growth stage i

In the program, daily average soil moisture and salinity statuses in the root zone are computed using volumetric balances. These are converted to matric and solute suctions and added to get total daily average integrated soil moisture suctions. Daily evapotranspiration rates are computed using an input water availability curve which relates the daily average root zone suction to the ratio of actual and maximum evapotranspiration. Yield reductions are then estimated using a linear production function of the form:

$$(1-\frac{Ya}{Ym}) = \sum_{i=1}^{n} k_{y_i} \; (1-\frac{ETa_i}{ETm_i}) \tag{7}$$

where

Y_a = actual harvested yield

Y_m = maximum attainable harvested yield

k_{y_i} = a response factor defined for each growth stage

ETa_i = actual evapotranspiration in growth stage i

ETm_i = maximum evapotranspiration in growth stage i

n = number of growth stages

This formulation combines soil moisture and salinity effects on plant growth by assuming that the matric and solute suctions are physically and physiologically additive and yield reduction is due to water stress. Further, it accounts for different crop responses at different growth stages and also for the variation in root depth with time. High water table effects are incorporated indirectly by considering their effect on root depths and the contribution of salts to the root zone from the water table. If the water table is at a level sufficiently below the root zone so that root elongation and respiration are not impeded and upflow of saline groundwater is negligible, yields will be a function of the frequency and quantity of irrigation, soil properties, and initial root zone conditions. However, if the water table or the saturated region above the water table infringes on the root zone, water uptake is assumed to cease in the saturated region. Further, if the saturated region remains in the root zone for an extended period of time, the roots in this region are killed. Crop growth will subsequently be affected by the diminished depth of transpiring roots.

EXAMPLE RESULTS

Data gathered in the Beni Magdoul area of Middle Egypt has been used to illustrate the program capabilities (Durnford 1982). Beni Magdoul is located in a very flat flood plain in the Nile delta. Precipitation occurs, on average, during only five days per year with an average yearly accumulation of less than 3 cm. The soils are alluvial deposits from the Nile suspended matter with clay contents ranging from 40 to 60 percent and occasionally higher. The area has a high water table which is generally within a meter of the ground surface. The irrigation water is very good quality but soil salinity is sufficiently high at times during the year to cause adverse effects on crops grown in the area.

The cropping intensity in Beni Magdoul is approximately two with corn, vegetables and Berseem clover as primary crops. The program has been used to investigate the two-crop season at Beni Magdoul. Results obtained showed that drain installation is not economically feasible based on the short-term analysis made in the computer model. However, the program results also indicated that a long-term deterioration of the groundwater quality may be occurring and drain installation may be economically feasible to prevent a long-term decline in the productivity of the area.

Program output includes daily and seasonal evapotranspiration values, net water added to the water table at each irrigation, the quantity of upflow from the water table to the root zone, and plots of daily averaged soil moisture and salinity levels in the root zone, the water table hydrograph and the effective root depths. Figures 3 and 4 are examples of the final output obtained. The curves shown are annual costs and yields obtained for the summer

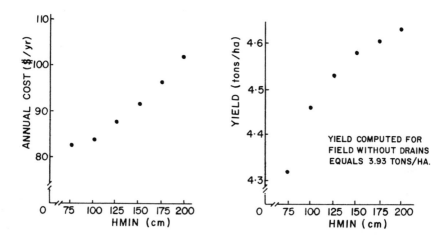

Fig. 3   Drain Costs   Fig. 4   Crop Yields

crop, corn, in 1978. In this case, the yield without drains was estimated to be 3.93 metric tons/hectare. The maximum attainable yield input was 4.75 tons/hectare. As these figures show, for the cost structure and input data assumed, the incremental cost of installing drains at levels greater than HMIN = 125 cm is significantly higher than at lower levels of protection. The corresponding yields, however, begin to level off after HMIN = 150 cm. In other cases investigated, yields actually decreased at higher levels. This decrease is because soil moisture levels decreased with increasing HMIN. By assigning a dollar value to the increase in yields and comparing curves such as those shown in Figs. 3 and 4, an optimal drain system for the data input can be determined.

SUMMARY AND CONCLUSIONS

A methodology to determine an economically optimum subsurface drainage system in irrigated areas is presented. The method combines a cost minimization procedure with an incremental net benefit search. The optimization routine eliminates the necessity of evaluating a large number of drain spacing and depth combinations and makes evaluation of the effects of particular variables on the results more traceable. The benefit analysis used attempts to estimate benefits in terms of those variables which are often the primary reasons for drain installation in arid regions.

REFERENCES

1. Bresler, E. and D. Yaron. 1972. Soil water regime in economic evaluation of salinity in irrigation. Water Resources Research 8(4): 791-800.

2. Bouwer, H. 1974. Developing drainage design criteria. In: van Schilfgaarde, J. (ed.). Drainage for agriculture. Agronomy 17: 67-79. American Society of Agronomy. Madison, WI.

3. Durnford, D. 1982. Optimal Drain Design: An Incremental Net Benefit Approach. Ph.D. Dissertation. Colorado State University, Fort Collins, CO. 269 p.

4. Hiler, E. A. 1977. Drainage requirements of crops. Third National Drainage Symposium. ASAE. St. Joseph, MI. 127-129.

5. McWhorter, D. B. 1977. Drain spacing based on dynamic equilibrium. J. Irrig. and Drain. Div., ASCE 103(IR2): 259-271.

6. Moody, W. T. 1966. Nonlinear differential equation of drain spacing. Proceedings, ASCE 92(IR2): 1-9.

7. Wadleigh, C. H. and A. D. Ayers. 1945. Growth and biochemical composition of bean plants as conditioned by soil moisture tension and salt concentration. Plant Physiology 20: 106-132.

OPTIMIZING DRAINAGE SYSTEM DESIGN FOR CORN

R. W. Skaggs   A. Nassehzadeh-Tabrizi
Member ASAE   Member

Soil water, either too much or too little, is the single most limiting
factor for crop production in North Carolina. This is especially true in
the Coastal Plains and Tidewater Regions where both excessive and
deficient soil water conditions frequently occur. Growers estimate that
cumulative water related losses amount to as much as a total crop failure
in one out of every three years. Artificial drainage is essential for crop
production on about 35% (1.0 million hectares) of the total North Carolina
cropland. Drainage is necessary to provide trafficable conditions for
seedbed preparation, harvesting and other field operations, and to remove
excessive soil water to insure a suitable environment for the growing crop.

Artificial drainage systems may include open ditches or buried drain tubes
to provide subsurface drainage and surface grading or land forming for
surface drainage. It has been recognized for a long time (van Schilfgaarde,
1965) that artificial drainage systems should be fitted to the soils, crops
and climatological conditions. However, methods for selecting the optimum
combination of surface and subsurface drainage for given soil and site
conditions have not been available in the past.

Methods based on the computer simulation model, DRAINMOD, have been
developed recently to analyze the effects of drainage systems on soil water
conditions and to predict their effects on annual and long-term average corn
yields (Skaggs, 1978; Hardjoamidjojo, et al. 1982). The purpose of this
paper is to use the model to analyze the effect of surface and subsurface
drainage intensities on corn yields over a 26-year period in eastern N.C.
Economic analyses are conducted for the poorly drained Rains sandy loam soil
to determine the optimum drainage system design for corn production.

THE MODEL

The simulation model, DRAINMOD, was described in detail by Skaggs (1978,
1980). It was developed for design and evaluation of multicomponent water
management systems which may include facilities for subsurface drainage,
surface drainage, subirrigation and sprinkler irrigation. The model is
based on a water balance in the soil profile. Input data include soil properties, crop parameters, drainage system parameters and climatological
data. The model simulates the performance of a given drainage system design
over a long period of climatological record. It predicts hourly values of
infiltration, drainage, surface runoff, evapotranspiration, water table
position, and factors such as the SEW_{30} (Wesseling, 1974) to quantify stress
due to excessive and deficient soil water conditions. DRAINMOD was modified
(Skaggs et al., 1979; Hardjoamidjojo and Skaggs, 1982) to include stress-
day-index models for estimating relative yield for corn as affected by excessive soil water conditions, deficient or drought conditions, and planting date delay.

The authors are: R. W. SKAGGS, Professor, and A. NASSEHZADEH-TABRIZI,
Research Associate, Biological and Agricultural Engineering Dept., N. C.
State University.

Simulations were conducted for several alternative drainage designs on a
Rains sandy loam (fine-loamy, siliceous, thermic typic Paleaqualt). The
Rains soil has poor surface drainage in the natural state with depression
storage of about 25 mm. It is underlain at a depth of 1.4 m by a heavy subsoil that restricts vertical water movement. Properties of the soil were
given in detail by Skaggs (1978) and are summarized in Table 1.

Table 1. Summary of Soil Property and Drainage System Parameter Inputs
for the Rains Sandy Loam Soil Considered

1. **Soil properties**
   Depth to restricting layer: 1.4 m
   Saturated hydraulic conductivity: Depth $\leq$ 1.1 m, 1.03 m/d

                                                1.1 m < Depth < 1.4 m, 0.24 m/d

   Water content at lower limit available to plants: 0.09 cm^3/cm^3
   Saturated water content in root zone: 0.37 cm^3/cm^3
   Required drainage volume for field work: 3.9 cm
   Minimum daily rainfall to stop field work: 1.2 cm
   Time after rain before work can resume: 2 days
   Green-Ampt Infiltration Parameters:

Water table depth (cm)	$A(cm^2/hr)$	B(cm/hr)
0	0.0	0.0
50	1.2	1.0
100	3.3	1.0
150	6.0	1.0
200	9.2	1.0
500	25.0	1.0

2. **Drainage system parameters:**
   Drain depth   1.0 m
   Drain diameter   100 mm
   Effect drain radius   5.1 mm
   Surface depressional storage*   2.5 mm, 25 mm
   Drain coefficient (as limited by hydraulics)   2.5 cm/day
   Drain spacing   5,10,15,22,30,45,60,
                                                        100,150,300 m

3. **Crop parameters (for corn)**
   Desired planting date   not later than April 15
                                                        (day 105)

   Working days required for seedbed preparation
     and planting   8.0 days
   Length of growing season   130 days
   Maximum effective rooting depth   30 cm

*Multiple entries indicate the different values used in the simulations.

Simulations were conducted for the 26 year period from 1950 to 1975 using
climatological data from Wilson, N.C. Climatological data from several
other N.C. locations also were used to determine the effects of in-state
weather variations on drainage designs required for optimum production.
The performance of drainage systems with drain spacings varying from 5
to 300 m and surface depressional storage depths of s = 25 mm (poor surface
drainage) and s = 2.5 mm (good surface drainage) were simulated. Seedbed
preparation was assumed to begin on March 15 (day 85) and 8 trafficable
days were required to plant the corn. If 8 days suitable for seedbed preparation occurred before April 10 (day 100), the corn was assumed planted
on day 100. If the 8th day occurred after day 100 the corn was assumed
planted on that day. It is recognized that the actual length of time
required for the planting operation depends on size of operation, equipment
and labor available, and many other factors. A period of eight days was
chosen arbitrarily as being typical for an eastern N.C. operation.

RESULTS

Average predicted relative yields for the 26 year simulation period are
plotted as a function of drain spacing in Figure 1. The relative yield is
defined as YR = Y/Y_0, where Y_0 is the base or potential yield that would be
obtained if there were no soil water stresses and the crop is planted on

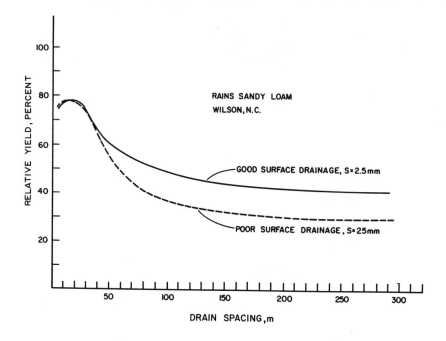

Fig. 1   Effect of Drain Spacing on Long-Term Average Relative Yields for
Both Good and Poor Surface Drainage on a Rains Sandy Loam

time and Y is the actual yield for a given year. Relationships for YR are
plotted for both good (depressional storage, s = 2.5 mm) and poor (s = 25
mm) surface drainage. Results in Fig. 1 show that a maximum average
relative yield (YR) of 0.78 is predicted for a drain spacing of L = 20 m
for good surface drainage and at L = 17 m for poor surface drainage. Higher
average yields were not obtained because of deficient soil water conditions
which caused drought stresses during several years. Yields increased with
better drainage (smaller drain spacing) until the maximum was obtained.
Further decreases in drain spacing caused the average YR values to drop
slightly showing a tendency for over-drainage if the drains are placed too
close together.

The results show clearly that the drain spacing required for a given level
of production is dependent on the quality of surface drainage. The effect
of surface drainage increases with drain spacing and is particularly
important for poor subsurface drainage, i.e. drain spacing greater than
about 75 m. For example, the predicted relative yield for a drain spacing
of 100 m, which is a typical spacing for open ditch drains in eastern N.C.,
is YR = 0.48 for good surface drainage versus YR = 0.37 for poor surface
drainage. Assuming an average potential yield of 11,000 kg/ha (175 bu/ac),
the predicted average yields would be 5300 kg/ha (84 bu/ac) and 4000 kg/ha
(64 bu/ac), for good and poor surface drainage, respectively.

It is probably obvious to the reader that the relationships given in Fig. 1
will not hold for all Rains sandy loam soils. The hydraulic conductivity,
depth to restrictive layer, soil water characteristics, and other soil
properties on which drainage depends may vary widely from site to site
within a given soil type. In most cases on-site measurements of the soil
property inputs should be made and used in DRAINMOD to optimize the drainage system design.

The effect of differences in hydraulic conductivity on the response of relative yield to drain spacing can be estimated by using the data given in the SCS soil survey interpretations. Relative yield-drainage spacing relationships obtained by using the upper and lower limits for K as given by the SCS data are plotted in Fig. 2. The relationship given in Fig. 1 also is plotted for comparison. The drain spacing required to maximize relative yield varies from 10 m for the lower limit K to 26 m for the upper limit K.

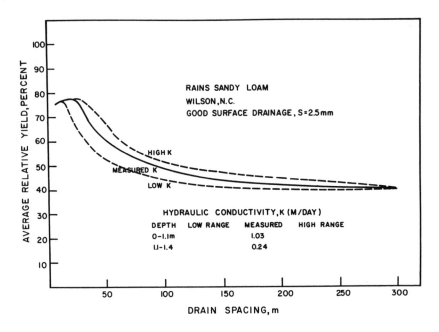

Fig. 2 Relative Yield – Drain Spacing Relationships for the Range of Hydraulic Conductivity (K) values given in the SCS Soil Survey Interpretation Data for the Rains Series. The Relationship for the Measured K Value is Given for Comparison

The effects of location on the relationship between average relative yield and drain spacing is shown in Fig. 3. Simulations were conducted for a 15 year period from 1956 to 1970 using climatological data from six eastern North Carolina locations: Elizabeth City, Greenville, Wilson, Raleigh, Wilmington and Laurinburg. The 15 year period was chosen because all stations had a complete weather record for that period. Differences in weather among the locations caused some differences in the predicted relative yield-drain spacing relationship. However the relationship had a similar form for all locations and the maximum predicted relative yield occurred for drain spacings between 15 and 20 m in all cases. Maximum predicted yields for the Wilmington station were about 10 percent higher than for the other locations. This difference is primarily due to more rainfall and less drought stress during June and July than recorded at the other stations. However yield response for poorly drained conditions (large drain spacings) for Wilmington was very similar to the other locations.

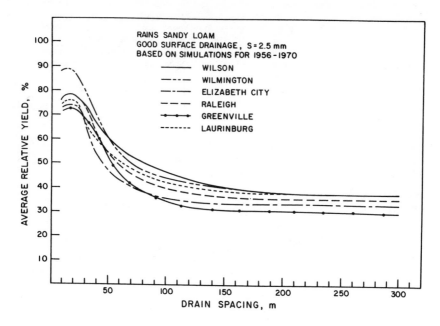

Fig. 3 Predicted Average Relative Yield Response to Drain Spacing
for Six Eastern N.C. Locations. Surface Drainage for the
Rains Sandy Loam was Assumed to be Good with Depressional
Surface Storage of Only 2.5 mm

ECONOMIC ANALYSIS

Whether a farmer can afford to install a given artificial drainage system
depends on its effect on profits. A drainage system can represent a sizable
investment and it is possible that the design needed to maximize yields may
be too expensive to maximize profits. An economic analysis was conducted to
determine the effect of drain spacing and surface drainage on long-term
average profits for corn. Analyses were conducted for both the Portsmouth
and Rains soils.

Income

The annual income for a given drainage system was calculated as the predicted average relative yield times the base or potential yield times the
assumed unit crop values. The potential yield for the Rains soils is
estimated as 11,000 kg/ha (175 bu/ac). As an example, consider a 20 m
drain spacing for the Rains soil with good surface drainage. From Fig. 1
the average relative yield is 78% so, on the average, we could expect a
yield of 0.78 x 11,000 = 8600 kg/ha (136 bu/ac). Then assuming an average
price of $0.118/kg ($3.00/bu) gives an annual income of $1010/ha ($409/ac).
Values for predicted average annual income for a range of drain spacings are
given in Table 2 for both good and poor surface drainage.

Table 2. Predicted Average Annual Income as Affected by Drain Spacing and Surface Drainage on the Rains Soil

Drain spacing	Relative yield		Actual yield*		Income**	
	Good surface drainage	Poor surface drainage	Good surface drainage	Poor surface drainage	Good surface drainage	Poor surface drainage
5 m	0.75	0.73	8250	8030	$970/ha	$944/ha
10	0.76	0.76	8360	8360	983	983
15	0.77	0.78	8470	8580	996	1002
20	0.78	0.77	8580	8470	1010	996
25	0.77	0.76	8470	8360	989	976
30	0.75	0.73	8250	8030	970	944
40	0.65	0.65	7150	7150	840	840
50	0.60	0.55	6600	6050	776	711
60	0.56	0.49	6160	5390	724	633
80	0.51	0.41	5610	4510	659	530
100	0.48	0.37	5280	4070	620	478
150	0.43	0.31	4730	3410	556	401
200	0.42	0.30	4620	3300	543	388
300	0.40	0.30	4400	3300	517	388

*
Assuming a potential yield of 11,000 kg/ha (175 bu/ac).
**
Assuming a corn price of $0.118/kg (3.00/bu).

Costs

Two types of costs were considered in the economic analysis: annual drainage system costs which included the initial cost amortized over its estimated useful life plus an allowance for maintenance, and normal production costs for corn. Production costs for corn in North Carolina, which included total fixed and variable costs but excluded land cost, were estimated as $483/ha from extension material prepared by the Department of Economics and Business at North Carolina State University.

The cost of subsurface drainage per unit area depends on the size and shape of the land area to be drained and site parameters such as the existence of an adequate gravity outlet, old field ditches or canals. The subsurface component was assumed to be made up of 10-cm (4-inch) corrugated plastic tubing emptying into existing drainage canals. Drainage costs were estimated based on information supplied by North Carolina drainage contractors assuming an installation size (area) of at least 40 ha (100 ac.) and typical field conditions for eastern North Carolina. The drainage system was assumed to have a useful life of 30 years. Maintenance costs were estimated at $20/ha/yr ($8/ac/yr) for surface drainage and two percent of the annual amortized cost for subsurface drainage. Total annual costs, including both drainage system and production costs for corn, are tabulated for a range of drain spacings in Table 3.

Profit

Annual profit was calculated by subtracting the estimated annual costs (Table 3) from the predicted annual income (Table 2). Predicted annual income, costs and profit are plotted as functions of drain spacing in Fig. 4 for the Rains soil with good surface drainage. The effect of drain spacing on profit is shown in Fig. 5 for both good and poor surface drainage. These results show that maximum profit of about $400/ha will be obtained for drain spacing of about 24 m without improved surface drainage (Fig. 5). For good surface drainage, a somewhat smaller maximum average profit of $360/ha will be obtained at a drain spacing of 27 m. The calculated maximum profit was higher for poor than for good surface drainage because it simply cost more to provide and maintain good surface drainage (Table 3) than to place the subsurface drains closer together to obtain nearly the same yield (Fig. 1).

Table 3. Estimated Total Annual Cost[*] for the Production of Corn as Affected by Drain Spacing and Surface Drainage (s.d.) Treatment

Drain spacing	Drain length per hectare	Initial cost		Annual drainage system cost[**]		Annual maintenance cost		Annual production cost	Total annual cost	
		Good s.d.	Poor s.d.	Good s.d.	Poor s.d.	Good s.d.	Poor s.d.		Good s.d.	Poor s.d.
5 m	2000 m	$4247/ha	$4000/ha	$527/ha	$497/ha	$30/ha	$10/ha	$483/ha	$1040/ha	$990/ha
10	1000	2247	2000	279	248	25	5	483	787	736
15	667	1581	1334	196	166	24	4	483	703	653
20	500	1247	1000	155	124	23	3	483	660	610
25	400	1047	800	130	99	22	2	483	635	584
30	333	913	667	113	83	22	2	483	618	568
40	250	747	500	93	62	21	1	483	597	546
50	200	647	400	80	50	21	1	483	584	534
60	167	581	334	72	41	21	1	483	576	525
80	125	497	250	62	31	21	1	483	566	515
100	100	447	200	56	25	21	1	483	560	509
150	66.7	380	133	47	17	20	0	483	550	500
200	50	347	100	43	12	20	0	483	546	495
300	33	313	66	39	8	20	0	483	542	491

[*]All costs rounded to nearest 1 dollar. $/ha can be converted to $/ac. by multiplying by the factor 0.405.

[**]Assuming initial cost amortized over 30 years at 12% interest.

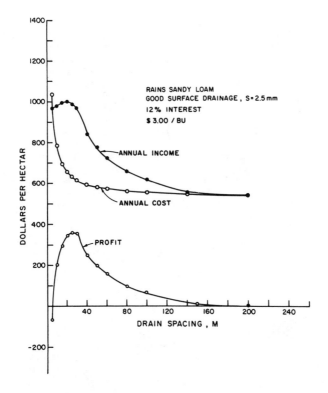

Fig. 4 Effect of Drain Spacing on Annual Income, Cost, and Profit for a Rains Sandy Loam Soil with Good Surface Drainage

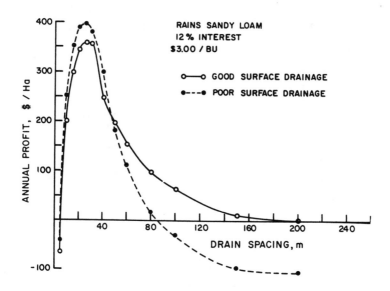

Fig. 5 Predicted Annual Profit from Production of Corn on a Rains
Sandy Loam Soil as Affected by Drain Spacing and Surface
Drainage

The need for good surface drainage is crucial when subsurface drainage is
poor. For example, consider a drain spacing of 100 m, a typical spacing for
open ditch drainage in eastern N.C. The average profit for a 100 m spacing
on this Rains soil (Fig. 5) would be $60/ha for good surface drainage com-
pared to a loss of $31/ha for poor surface drainage. This need is recog-
nized by most farmers who usually shape the fields in a turtle-back fashion
and plant on beds to improve surface drainge. However, our results show
that the use of subsurface drain tubing spaced about 25 m apart would
greatly increase profits for this Rains soil.

The results given in Figs. 4 and 5 assume a price for corn of $0.118/kg
(3.00/bu). The effect of both higher and lower prices on profits is shown
in Fig. 6 for poor surface drainage. Obviously, lower prices reduce profits. However, the drain spacing required to maximize profits was not
sensitive to price, varying only from 25 m to 23 m for corn prices ranging
from $2.00/bu to $4.00/bu.

Another assumption made in calculating profits was that the potential
yield, the average yield that would be obtained under perfect soil water
conditions or in the absence of soil water stresses, is 11,000 kg/ha (175
bu/ac). This potential yield may be too high for some Rains sandy loams and
too low for others, depending on the soil fertility, management practices,
problems with weeds or insects, and other factors not directly related to
soil water conditions. Because yearly gross income is simply the product of
price and yield, the effects of different potential yields on profit can
also be obtained from Fig. 1. The potential incomes corresponding to the
various prices are listed on Fig. 6. The corresponding profit curve for a
given potential income are valid regardless of the combination of price and
potential yield that results in that income. For example, a potential yield
of 175 bu/ac at a price of $2.50/bu would result in a potential income of
175 bu/ac x $2.50/bu x 2.47 ac/ha = $1080/ha. This is the same potential

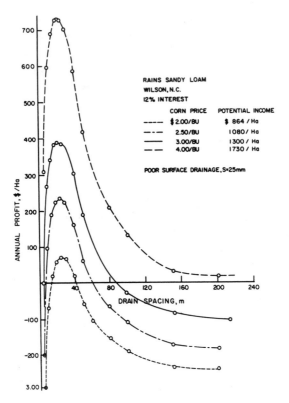

Fig. 6 Effect of Corn Price and/or Potential Yield on the
Relationship Between Average Annual Profit and Drain
Spacing for the Rains Soil. Note that Potential
Income = Price x Potential Yield

income that would result from a potential yield of 146 bu/ac at a price of
$3.00/bu, 146 bu/ac x $3.00/bu x 2.47 ac/ha = $1080/ha. So the profit curve
for a potential income of $1080/ha is valid for a potential yield of 175
bu/ac at $2.50/bu as well as for 146 bu/ac at $3.00/ha. Again it may be
concluded that the optimum spacing is not sensitive to either the potential
yield or price for the range of conditions considered. Obviously, the price
and potential yield have large effects on profits, but they have only small
effects on the drainage design required to maximize profits.

Economic analyses were conducted for the relative yield – drain spacing
relationships given in Fig. 3 to determine the effect of location on the
optimum drain spacing and on profit. Results for the six locations are
summarized in Table 4. Although maximum annual profits varied from $290 to
$471 per hectare per year, the drain spacing required to maximize profits
varied from only 24 m to 27 m. This indicates that climatological data
from rather widely scattered stations can be used for drainage design in
eastern North Carolina. Data from the nearest station that has the
required hourly rainfall and temperature records should be used. But, if a
nearby station does not have the required length of record, data from a more
remote location can be used to select a drainage system design to optimize
profits.

Table 4. Average Annual Profit and the Corresponding Drain Spacings
          Required to Maximize Profit for the Rains Sandy Loam at Six
          Eastern N.C. Locations

Location	Average annual profit	Drain spacing for maximum profit
Elizabeth City	$309/ha	25 m
Greenville	290	25
Laurinburg	331	24
Raleigh	361	25
Wilmington	471	24
Wilson	355	27

The annual cost of the drainage system depends on the useful life of the
system and the interest rate used for amortization. A 12% interest rate
was assumed in the economic analyses discussed previously in this paper.
The effect of interest rate on profit and on the drainage spacing required
to maximize profit is shown in Table 5 for the Rains soil with poor surface
drainage. While annual profits decrease with increase in interest rate, the
drain spacing required to maximize profit remained about the same for the
range considered.

Table 5. Effect of the Interest Rate Used in Amortizing Costs of the
          Drainage System on Annual Profit and on Drain Spacing Required
          to Maximize Profit for the Rains Sandy Loam with Poor Surface
          Drainage

Interest rate	Optimum drain spacing	Annual profit
10%	24 m	$410/ha
12%	24	395
14%	25	380

Although the useful life of the drainage system easily may exceed the 30
years assumed, it is unlikely that a farmer could borrow the initial
capital requirements for that length of time. The farmer's ability to
borrow money for drainage and/or his willingness to invest in drainage
depends on the length of time required for the investment to pay for itself.
The time required for return of investment depends on the size of the
initial investment and the increase in profits due to drainage. Results in
Fig. 5 show that, on the average, corn cannot be grown at a profit on this
soil without artificial drainage. A drainage system consisting of good
surface drainage and subsurface drains placed 27 m apart would have an
initial cost of $1000/ha (Table 3). Total annual costs would amount to
$628/ha (Table 3), of which $124/ha is allocated for amortizing the initial
investment over the 30 year life of the drainage system. Net profit is
predicted to be $360/ha. Assuming that all profits are used to pay off the
initial investment, an average total of $360 + $124 = $484/ha would be
available to pay on the debt each year. This would amount to an average
return on the initial investment of $484/1000 = 48.4%, and the drainage
system would pay for itself in only three crop years. The times required
to pay off the initial investment and the average annual return on investment are summarized in Table 6 for a range of corn prices. The calculated
return on investment may be somewhat misleading because it assumes that all
profits are derived from the drainage investment. The investment in land is

Table 6. Average Annual Return and Number of Years Required to Pay Off
Initial Investment for Drainage Systems on a Rains Sandy Loam Soil

Surface drainage	Initial drainage system cost	$3.00/bu			$2.50/bu			$2.00/bu		
		Annual return	% on invest-ment	Payoff time	Annual return	% on invest-ment	Payoff time	Annual return	% on invest-ment	Payoff time
Good, s = 2.5 mm	$1000/ha	$484/ha	48	3 yrs.	$318/ha	32	4 yrs	$157/ha	16	13 yrs
Poor, s = 25 mm	840	499	59	2	339	40	3	174	21	8

not considered. However, corn could not be grown at a profit on these soils
without artificial drainage (Fig. 6). So investment in an optimum
drainage system would increase the average annual return by the amounts
shown in Table 6.

SUMMARY AND CONCLUSIONS

The simulation model DRAINMOD as modified to estimate the effects of soil
water stresses on corn yields was used to determine the effect of surface
and subsurface drainage on corn production for a Rains sandy loam in
eastern North Carolina. Simulations were conducted for both good and poor
surface drainage and for subsurface drain spacings ranging from 5 m to
300 m. Economic analyses were made to identify the drainage alternatives
needed to maximize profits from corn production on the two soils.

The following conclusions are based on the results of the simulations and
analyses.

1. Surface and subsurface drainage have a significant effect on long-term
    average yields for the Rains sandy loam soil investigated. Predicted
    average yields could be increased by 63% by providing better subsurface
    drainage than that given by conventional drainage ditches spaced 100 m
    apart. Increases were even greater for conditions of poor surface
    drainage.

2. Drains should be spaced 20 m apart for good surface drainage and 17 m
    apart when the surface drainage is poor to maximize long-term average
    yields for the soil investigated. Profits are maximized at somewhat
    larger drain spacings than required to maximize yields. Maximum
    profits were predicted for spacings of 24 m to 27 m depending on the
    surface drainage intensity.

3. Results reported herein for the particular Rains sandy loam investigated
    are not valid for all soils within the series. Variation of saturated
    hydraulic conductivity from site to site could cause the drain spacing
    required to maximize yield to vary by more than a factor of 2.5. A
    drainage system design for a given site should be based on soil pro-
    perty measurements for that site.

4. Corn prices and interest rates have a predictably large effect on
    profit. However, the drainage design required to maximize profit was
    affected only slightly by these variables within the range considered.

REFERENCES

1. Hardjoamidjojo, S. and R. W. Skaggs. 1982. Predicting the effects of drainage systems on corn yields. Agricultural Water Management, Vol. 5:127-144, Elsevier Publ. Co., Amsterdam.

2. Hardjoamidjojo, S., R. W. Skaggs and G. O. Schwab. 1982. Predicting corn yield response to excessive soil water conditions. TRANS. of the ASAE, Vol. 25(4):922-927.

3. Schwab, G. O. 1976. Effect of drainage on corn, oat and soybean yields at North Central Branch OARDC 1962-1972. The Ohio Report, March-April 1976, Ohio Research and Development Center.

4. Shaw, R. H. 1974. A weighted moisture stress index for corn in Iowa. Iowa State Journal of Research, Vol. 49(2):101-114.

5. Shaw, R. H. 1976. Moisture stress effects on corn in Iowa. Iowa State Journal of Research, Vol. 50(4):335-343.

6. Skaggs, R. W. 1978. A water management model for shallow water table soils. Report No. 134, Water Resources Research Institute of the University of North Carolina, N. C. State University, Raleigh, 178 pp.

7. Skaggs, R. W. 1980. A water management model for artificially drained soils. Tech. Bulletin No. 267. North Carolina Agricultural Research Service, N. C. State University, Raleigh. 54 pp.

8. Skaggs, R. W., S. Hardjoamidjojo, E. H. Wiser and E. A. Hiler. 1979. Simulation of crop response to surface and subsurface drainage systems. TRANS. of the ASAE, In Press.

9. van Schilfgaarde, Jan. 1965. Transient design of drainage systems. J. Irrig. and Drain. Div., ASCE, Vol. 91(IR3):9-22.

10. Wesseling, J. 1974. Crop growth and wet soils. Ch. 2 in Drainage for Agriculture. J. van Schilfgaarde, ed., American Society of Agronomy, Madison, WI. pp. 7-37.

DRAINAGE GUIDE IMPROVEMENTS FOR BETTER
DRAINAGE DESIGN

W. J. Ochs   R. D. Wenberg   P. E. Lucas
Member ASAE   Member ASAE   Member ASAE

Drainage guides have been used for decades by the Soil Conservation Service (SCS) staff and others as a basic tool to design surface and subsurface drainage systems. The guides were developed mainly for the humid section of the United States and Canada.

These guides were usually organized into groups of wet soils (SCS, 1974) that generally required the same or similar drainage recommendations. These recommendations usually included surface treatment needs, drainage coefficients, side slopes and maximum velocities for ditches, depth of field ditches, depth and spacing of subsurface drains, and spacing of surface ditches if applicable.

A drainage guide facilitates efficient drainage system design. It requires a great deal of time and effort to update. Four recent technological advancements are helping to expedite the preparation of drainage guides, improve recommendations, and reduce the cost of preparing or updating them. The advancements are:

1. "DRREC"--A computerized system for storage and retrieval of existing drainage recommendations and for the retrieval of associated detailed soils data from other computerized files (Wenberg, 1979).

2. "SOILSORT"--A computer program devised for use by SCS to utilize information obtained by the Cooperative Soil Survey (SCS, 1974; Lucas, 1981). With this program, the user can obtain pertinent information from the soil interpretation records, as well as from specialists, to sort soils into drainage recommendation groups. Information regarding general drainage practices applicable to the group of soils and, if appropriate, the theoretical subsurface drain depth and spacing can also be obtained.

3. "DRAINMOD"--A computer model developed to provide accurate and specific information on the soil-water-plant system's response to individual drainage designs and recorded weather data (Skaggs 1975, 1978, and 1980).

4. "Drainability Studies"--Field programs to determine the drainability of soils that have shallow restrictive layers and for heavy clay soils (Patronsky and Schwab 1979). Past drainage recommendations for these soils have not been very accurate. Much more information has been obtained now, and sound recommendations can be made for many of these soil areas.

The authors are: Walter J. Ochs, National Drainage Engineer, U.S. Department of Agriculture, Soil Conservation Service, Washington, D.C.; Richard D. Wenberg, Drainage Engineer, South National Technical Center, USDA-Soil Conservation Service, Fort Worth, Texas; and Paul E. Lucas, Drainage Engineer, Midwest National Technical Center, USDA-Soil Conservation Service, Lincoln, Nebraska

These tools are being used by SCS to improve drainage guides for humid areas and to improve drainage system designs. The objective of this paper is to briefly describe these tools and encourage prompt revision of drainage guides to save farmers money.

DRAINAGE RECOMMENDATIONS

The drainage recommendation (DRREC) computer program (Wenberg 1979) is a system for storage and selective retrieval of drainage recommendations and soils data. The purpose of the system is to facilitate coordination of drainage design criteria and provide the available pertinent soils information for updating drainage guides. The program is intended for use in the humid section of the United States as presently developed.

All drainage recommendations are related to a specific soil series and may be refined for specific land use for both surface and subsurface drainage parameters. Relevant information from soil interpretation records, developed as a result of the Cooperative Soil Survey (SCS, 1974) and previously computerized, are related internally to the drainage recommendations. This information is referred to as "estimated soil properties." The main items used from individual soil interpretation records are soil texture and permeability by depth, flooding potential and duration, depth to water table and season, and depth to bedrock.

Printouts can be obtained from the DRREC program to facilitate review and revision as well as to provide sheets for direct printing of revised drainage guides. One printout provides an effective and efficient means of comparing drainage recommendations for an individual soil made by other states having the same soil (refer to example no. 1 at the end of this paper). Example no. 2 is a printout that could be inserted directly in a drainage guide.

This program has been used by at least 20 humid-area states and is expected to provide a great deal of help in future years. An evaluation of drainage coefficients by Wenberg (1980) was facilitated by using the DRREC program.

SOILSORT

The computer program for sorting soils, SOILSORT, was developed by Lucas (1981). It utilizes information in the soil interpretation records along with three drain-spacing formulas to sort soils into drainage recommendation groups. Supplementary information that facilitates the soil-sorting process is entered into the computer with other needed data. This program provides a means by which a state's entire list of soils can be recorded in less than a day. For this investment in time, the user takes advantage of the wealth of information obtained through the Cooperative Soil Survey. The SOILSORT program uses most of the same information from individual soil interpretation records as used in the DRREC program.

Drainage Formulas

The SOILSORT program can determine the drain depth and spacing relationship using the following steady-state drainage formulas:

1.) Ellipse Formula (SCS, 1971)—used if the distance between the drain and the barrier is 0.3 m (1 ft.) or less.

2.) Hooghoudt Formula (Hooghoudt, 1940)—used for most soils and is particularly applicable to drains in homogeneous soil.

3.) Ernst Formula (Ernst, 1954)--used for soils that have two layers with
     differing permeabilities, where the permeability of the upper layer is
     less than that of the underlying layer.

SOILSORT Benefits

The SOILSORT program provides a relatively inexpensive and quick means of
establishing or checking drainage recommendation groups and calculating the
theoretical subsurface drain depth and spacing. It is valuable when used to
update drainage guides.

DRAINMOD

The DRAINMOD computer model, developed by Skaggs (1975, 1978, and 1980), provides a sound method for designing and evaluating drainage and other water
management systems on soils in humid areas. This model provides valuable
and accurate recommendations which improve the confidence level of drainage
guides. It also provides good design checks for improving recommendations
for drainage systems involving subsurface irrigation and drainage recommendations for waste water disposal sites.

The basis of this computer simulation model is a water balance at the midpoint between two parallel drains in soils that have a shallow water table
(a finite depth to the restrictive layer). Using the approach of successive
steady-state equilibrium, the unsteady and transient water movement in the
soil profile are analyzed as steady-state flow at sufficiently small time
increments. Each term in the water balance equation is individually evaluated from input data and brought into the water balance at the end of each
time period. If the proper climatological, physical property of the soil
and crop growing information are provided, the model is capable of determining
the hour-by-hour, day-by-day moisture fluctuations in the soil profile for a
defined water management scheme.

DRAINABILITY STUDIES

The field program described by Patronsky and Schwab (1979) provides field
procedures to evaluate drainability and develop recommendations for soils
with a shallow restrictive layer and heavy clay soils. These soils have
been a problem for many years, and drainage recommendations generally have
been limited to surface drainage. The soils are being drained now, using
combination surface and subsurface drainage systems. Recommendations in
drainage guides are being adjusted as investigations are completed to reflect
the new findings.

Approximately 15 soil series in Ohio, Indiana, Illinois, and surrounding
states have been investigated and plans include work on about 25 more during
the next 5 years. Reporting the findings (Schwab, et al., 1982) is a very
necessary ongoing effort.

SUMMARY AND CONCLUSIONS

Drainage guides provide recommendations that make drainage system designs
quicker, easier, and will improve them if guides are periodically updated
(Beauchamp and Fasken, 1954). The four technological advancements discussed
in this paper are valuable tools for improving and accelerating the revision
of drainage guides. The prompt revision of these guides will save farmers a

great deal of money by providing sound technical recommendations soon after new information becomes available.

The DRREC computer model is a basic tool for comparing recommendations, using existing soils data, and updating guides promptly as well as economically. When this program is used with the SOILSORT computer program, the theoretical drain depth and spacing along with other general recommendations can be compared and evaluated quickly without additional field work.

If recommendations presently included in the guides appear suspect or seem to be improvable, the DRAINMOD computer model and/or drainability studies can be used to determine accurate recommendations. The drainability studies are more applicable to soils that have a shallow restrictive layer or for heavy clay soils. DRAINMOD is mainly applicable to humid-area soils that have a high water table. DRAINMOD can be applied sooner and at considerably less cost than drainability studies.

Drainage guides should be updated and improved promptly when new information becomes available. They provide information that can save farmers money. The four tools described in this paper will facilitate the revision of drainage guides and provide the opportunity for periodically improving them.

REFERENCES

1. Beauchamp, K. H. and G. G. Fasken. 1954. Progress in preparing drainage recommendations by soil types. ASAE Pap. from 47th Annual Mtg. 9p.

2. Ernst, L. F. 1954. Het Berekenen van Stationaire Groundwaterstromingen, Welke in een Vertikaal Vlak Afgebeeld Kunnen Worden, Rapport Bodemk. Inst. T.N.O. Grainingen.

3. Hooghoudt, S. B. 1940. Bijdragen tot de Kennis van enige natuur Kundige grotheden van de grond nr. 7 Versl. Landb. Onderz. 46: 515-707. Alegemeene Landsdrukkerij, The Hague.

4. Lucas, P. E. 1981. SOILSORT - A computer program for sorting soils into drainage recommendation groups. ASAE Pap. No. 81-2539. 19p.

5. Patronsky, R. J. and G. O. Schwab. 1979. A plan for field evaluation of subsurface drains in heavy soils and soils with a restrictive layer (humid area). 8th Tech. Conf. of the U.S. Comm. on Irrigation, Drainage, and Flood Control. International Commission on Irrigation and Drainage, Denver, Colorado. 18p.

6. Schwab, G. O., D. W. Michener, and D. E. Kopcak. 1982. Spacings for subsurface drains in heavy soils. Research Bull. No. 1141. Ohio Agricultural Research and Development Center, Wooster, Ohio. 20p.

7. Skaggs, R. W. 1975. A water management model for high water table soils. ASAE Pap. No. 75-2524.

8. Skaggs, R. W. 1978. A water management model for shallow water table soils. Tech. Rep. No. 134. Water Resources Research Institute of the University of North Carolina, North Carolina State University, Raleigh, North Carolina. 116p.

9. Skaggs, R. W. 1980. A water management model for artificially drained soils. Tech. Bull. No. 2527. North Carolina Agric. Res. Serv. North Carolina State University, Raleigh, North Carolina. 54p.

10. Soil Conservation Service, United States Department of Agriculture. 1971. Drainage of agricultural land. Soil Conserv. Serv., U.D. Dept. of Agric., Washington, D.C. 430p.

11. Soil Conservation Service, United States Department of Agriculture. 1974. National soils handbook. Soil Conserv. Serv., U.S. Dept. of Agric., Washington, D.C.

12. Wenberg, R. D. 1979. Drainage guide and water table management. ASAE Pap. No. 79-2544. 10p.

13. Wenberg, R. D. 1980. Evaluation of coefficients for subsurface drainage used by the Soil Conservation Service in the humid United States. ASAE Pap. No. 80-2572.

12/03/81   NORTH CAROLINA   SERIES CARTECAY

DEPTH IN	(- - - - USDA TEXTURE - - - -) MOD TEXT	MOD TEXT		PERM IN/HR	BEDROCK IN
0-9	SL	FSL	LS	6.0-20	>60
0-9	L	SIL	SICL	2.0-6.0	
9-40	SL	FSL	L	2.0-6.0	
40-60	LS	S	SL	6.0-20	

FLOODING - - - - - (DURATION = BRIEF   MONTH = DEC-MAR
                    FREQUENCY = COMMON)
HIGH WATER TABLE - - - - (DEPTH-FT = 0.5-1.5 MONTHS = JAN-APR)

(- - - - - - - - RECOMMENDATIONS - - - - - - - - -)

STATE	REC. DATE	AVG. LAND SLOPE	CROP--LANDUSE	DRAIN COEF.	SIDESLOPE Z MIN MAX		DEPTH IN MIN REC		SPACING FT MIN MAX		FILTER	SURFACE TRTMNT.

(SURFACE DRAINAGE - OPEN DITCH)

STATE	REC. DATE	SLOPE	CROP--LANDUSE	DRAIN COEF.	SIDESLOPE MIN	SIDESLOPE MAX	DEPTH MIN	DEPTH REC	SPACING MIN	SPACING MAX	FILTER	SURFACE TRTMNT.
ALABAMA	8/79	0-2	CROPLAND	45	1	4	12	36				SMOOTH
GEORGIA	02/76		CROPLAND				12	40				SMOOTH
SOUTH CAROLINA	06/77	0-2	CROPLAND	45	.5	1	24					
SOUTH CAROLINA	06/77	0-2	CROPLAND	45	10	10						
ALABAMA	8/79	0-2	PASTURE	30	1	6	12	30				SMOOTH
SOUTH CAROLINA	06/77	0-2	PASTURE	30	4	10	10					
SOUTH CAROLINA	06/77	0-2	PASTURE	30	.5	1	24					

(SUB-SURFACE DRAIN)

STATE	REC. DATE	SLOPE	CROP--LANDUSE	DRAIN COEF.	DEPTH MIN	DEPTH REC	SPACING MIN	SPACING MAX	FILTER	SURFACE TRTMNT.
ALABAMA	8/79	0-2	CROPLAND	1/2	30	48	80	120	CHECK N	
SOUTH CAROLINA	06/77	0-2	CROPLAND	1/2	30	42	80	110	CHECK N	

STREAM OVERFLOW HAZARD

STATE	REC. DATE	CROP--LANDUSE	SPACING	SURFACE TRTMNT.
GEORGIA	02/76	PASTURE	40	080	SMOOTH

```
                          DRAINAGE GUIDE RECOMMENDATIONS
        11/24/81                    TENNESSEE
                                                        SERIES   BEASON

        ( - - - - - - - - - - - ESTIMATED SOIL PROPERTIES - - - - - - - - -)

        DEPTH        ( - - - - USDA TEXTURE - - - -)     PERM      BEDROCK
         IN        MOD TEXT  MOD TEXT  MOD TEXT         IN/HR      IN

        0-7           SIL      SICL                      0.6-2.0    >60
        7-60          SICL     SIC       C               0.2-0.6
        60-80         VAR
FLOODING - - - - - - -(DURATION  = V.BRIEF                    MONTH =  DEC-APR
                       FREQUENCY = FREQUENT              )
HIGH WATER TABLE - - - - - -(DEPTH-FT = 1.0-2.0 MONTH = DEC-APR)


(- - - - - - - - - - - - - - - RECOMMENDATIONS - - - - - - - - - - - - - -)

AVG.      CROP--LANDUSE      DRAIN   SIDESLOPE  DEPTH  SPACING  FILTER   SURFACE
LAND                         COEF.      Z        IN      FT              TRTMNT.
SLOPE                                MIN REC   MIN MAX  MIN MAX


        (SURFACE DRAINAGE - FIELD DITCH)
          CROPLAND              45       2  8   12  36
          PASTURE               30       2  8   12  36

        (SUB-SURFACE DRAIN)
          CROPLAND             3/8               30  48   80 120 NONE
```

FIELD TESTS OF SOME DRAIN TUBE ENVELOPE MATERIALS

Robert S. Broughton, Samuel Gameda and Winston Gibson
Member ASAE   Assoc.Member ASAE

An experimental subsurface drainage system as shown in Fig. 1 was installed in fine sandy loam soil on the farm of Arthur Valois, Notre Dame du Bon Conseil, Québec, in August 1976.

The purposes of the project were:

1. To evaluate the ability of various envelope materials to keep sand out of the drain tubes but allow a good entry of water into drain tubes which are surrounded by sandy structureless soil.

2. To measure the discharge of water and sand from the drain tubes.

3. To measure the entry resistance or head loss near the drains to determine whether there are significant differences associated with the various envelope materials.

4. To watch for other complicating factors such as the formation of microbial sludges.

MATERIALS IN THE FIELD TEST

The drain tubes and envelope materials installed in August]976 were:

1. Standard corrugated polyethylene drain tubing with 120 slots per meter; slot size 1.5 mm x 30 mm and no envelope, open area was 54 cm^2/m.

2. Tubing with 590 small slots per meter;slot size approximately 0.7 mm x 8 mm without envelope;open area 33 cm^2/m.

3. 20 g/m^2 Cerex (spun bonded nylon) on 100 mm diameter polyethylene, PE tube with slots approximately 1.5 mm x 30 mm, 120 slots per meter.

4. 27 g/m^2 Reemay (spun bonded polyester) on 100 mm diameter PE tube with slots approximately 1.5 mm x 30 mm, 120 slots per meter.

5. Knitted nylon on 100 mm diameter PE tube with slots approximately 1.5 mm x 30 mm, 120 slots per meter.

6. 119 g/m^2 Typar (spun bonded polypropylene) on 100 mm diameter PE tube with slots approximately 1.5 mm x 30 mm, 120 slots per meter.

7. Coarse concrete sand envelope around 80 mm diameter PE tube. Slots were

The authors are:Robert S. Broughton, Professor, Samuel Gameda, Research Associate, Department of Agricultural Engineering, Macdonald Campus of McGill University, 21,111 Lakeshore Road, Ste. Anne de Bellevue, P.Q. H9X 1C0 and Winston Gibson, Extension Engineer, Field Engineering Division, Ministry of Agriculture, Land and Fisheries, Centeno, Trinidad.

oval 0.8 mm x 4 mm and 510 slots per meter;open area 16 cm^2/m.

The envelope materials in 3 to 7 above had shown the most promise in laboratory tests reported by Broughton et al. (1976). The layout of the drains is given in Fig. 1.

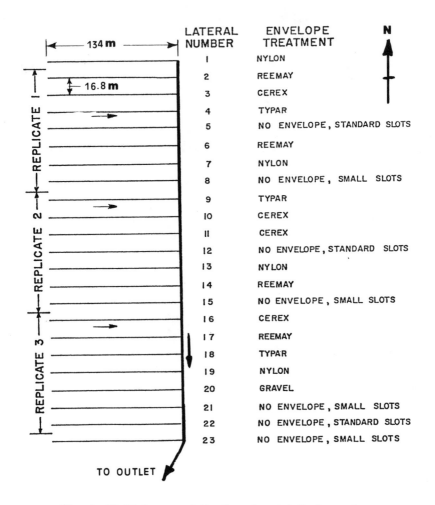

Fig. 1. Field Layout of the Experimental Drainage System

The field installation contained 23 laterals, 134 m long, spaced 16.75 m apart. There were three replicates of seven laterals which were enveloped as described above. The laterals were randomized within each replicate. Due to installation problems and costs, only one lateral was surrounded by a coarse concrete sand envelope. The remaining 2 laterals were used to install another replicate of the Reemay and the Cerex envelopes. An interceptor lateral was installed at each end of the experimental area. The out-flow from each lateral was measured at a manhole before it flowed into a collector drain which passed to an outlet. Two hundred and ninety-four observation pipes were placed to monitor the water table in the soil, the resistance to entrance of water into the drain tubes, the accumulation of sand in the drains, and the discharge of water and of sand from the laterals. The positions of the piezometers, the discharge observation chambers and the locations where the pipes were dug up to measure sediment are shown in Fig. 2. Typical particle size distribtuions for soil above and below the drains are given in Fig. 3.

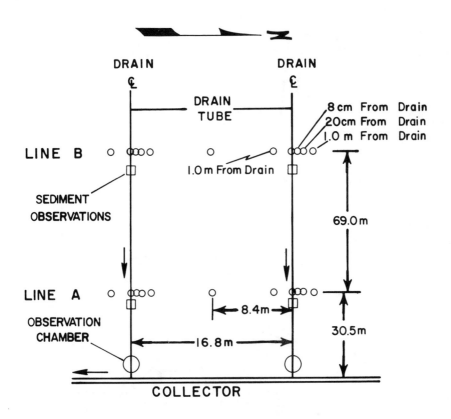

Fig. 2. Position of Piezometers, Sediment
and Discharge Observations

RESULTS AND DISCUSSION

Observations in the Spring of 1977

Snowmelt came early and gradually in the 1977 spring. This was followed by an abnormally low rainfall in April and May giving near drought conditions (about one-in-30 years' dryness). Thus, the drainage system did not get a severe test. The water table did not rise near to the soil surface at any time prior to August 1977.

Observations From July Through October 1977

In order to get results equivalent to a very wet season which would give envelope materials a severe test, a portable sprinkler irrigation system was installed.

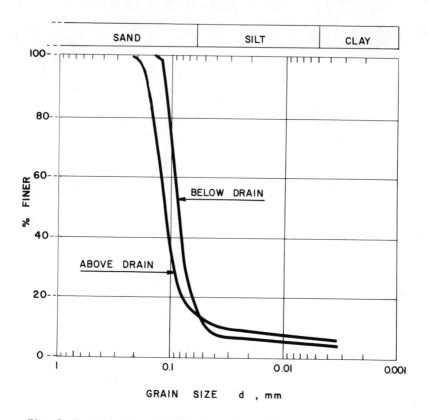

Fig. 3. Particle Size Distribution for Soil Samples Taken Above
and Below Lateral 4. Particle Size Distributions For Soil
Around Other Laterals Were Similar.

The replicates were irrigated in July and August 1977 to bring the water
table near the surface. Rainfalls in September and October also brought the
water table near the surface. Measurements were made of:
1) the water outflow rate from each lateral for several different water
   table heights,
2) the rate of fall of the water table,
3) the amount of sand coming out of each lateral,
4) the depth of sand which settled in each lateral,
5) the hydraulic conductivity of the soil in the field by
   a) the auger hole method
   b) the drain outflow method, and
   c) the falling water table method,
6) the drainable pore space vs soil water suction relationships for the
   subsoils,
7) the iron content of the soil,
8) the microbial growth outside and inside the drain tubes, and
9) the particle size distribution of sand found inside and outside the
   pipes.

Sediment in the Drain Tubes

Table 1 gives a summary of the depth of sand and microbial sludge observed
in the drain tubes in May and October 1977 and May 1978 at the 42 sampling
tees. In the cases where significant sand was found in the drain tubes at
the sampling tees, the tubes were also dug up upstream of the sampling tees

to observe the interior of the drain tubes.

Table 1. Observations of Total Depth* of Sand and Sludge in Drain Tubes, cm.

Treatment descriptions and dates of observations	Replicate 1		Replicate 2		Replicate 3	
	line A	line B	line A	line B	line A	line B
May 17-18, 1977						
Typar, 120 g/m^2	1.8	2.5	0.5	0.2	1.0	0.5
Reemay, 30 g/m^2	Negl.	0.5	1.4	3.8	3.1	2.0
Nylon sock	1.9	–	Negl.	Negl.	1.0	Negl.
Cerex, 20 g/m^2	Negl.	0.1	0.5	1.5	1.6	0.1
No envel. small slots	3.0[a]	1.8[a]	0.6[a]	0.6[a]	1.0[a]	0.8[a]
No envelope, Normal slots	4.0[a]	4.0[a]	2.9[a]	3.8[a]	1.7[a]	3.4[a]
Gravel and sand	–	–	–	–	0.6[b]	1.0[b]
October 7-8, 1977						
Typar, 120 g/m^2	0.8	0.8	0.7	0.5	1.1	0.8
Reemay, 30 g/m^2	1.0	0.6	1.1	1.2	0.9	2.9
Nylon sock	0.9	1.0	0.6	1.1	1.3	Negl.
Cerex, 20 g/m^2	0.1	0.2	0.8	1.2	0.9	0.5
No envel. small slots	4.3[a]	3.0[a]	1.2[a]	1.0[a]	1.8[a]	1.8[a]
No envelope, Normal slots	3.7[a]	5.0[a]	4.3[a]	8.2[a]	3.3[a]	4.3[a]
Gravel and sand	–	–	–	–	0.8[b]	1.2[b]
May 31, 1978						
Typar, 120 g/m^2	0.5	1.5	0.8	1.0	1.3	1.9
Reemay, 30 g/m^2	4.5	2.3	1.8	2.0	Negl.	3.1
Nylon sock	0.9	1.1	0.6	1.7	1.8	4.0
Cerex, 20 g/m^2	0.5	0.3	1.0	1.9	1.0	1.2
No envel. small slots	3.5[a]	3.0[a]	1.5[a]	1.9[a]	–	–
No envelope, Normal slots	6.0[a]	4.7[a]	7.5[a]	9.0[a]	2.0[a]	4.5[a]
Gravel and sand	–	–	–	–	1.0[b]	1.2[b]

[a] Mainly sand
[b] Sand and gravel
*Uniform depth of sand and/or sludge above the bottom of the tube
Negl: Negligible
– : No envelope in replicate (gravel and sand) or data unavailable

From Table 1 it can be seen that there was significant sand in the drains
without an envelope. The tube with normal slots in replicate 2 was
practically filled. The tubing with slot width 0.7 mm and length 8 mm had
less sand than the tubing with normal slots and no envelope. However,
there was a significant amount of sand in these tubes. It was noticed that
the amount of sand in the tubes increased from May 1977 to October 1977 to
May 1978.

It appears that sand is brought into the drain tubes as water flows into the
tubes. Also, the greater the drain flow, the greater the amount of sand is
brought into the tubes. A series of wet periods, perhaps requiring 4 to 6
years eventually will bring in enough sand to block drain tubes which have
no envelope.

A small amount of sand was found in the one drain tube which had the coarse
concrete sand envelope. This indicates that the particle size range of the
envelope material was not quite suitable to act as a good envelope for this

tubing.

The material in the tubes which were wrapped with a thin envelope fabric was mostly microbial sludge. It was noted that some soil particles were trapped within the microbial sludge.

There was almost no sand in the tubes which were wrapped with the 119 g/m^2 Typar and the nylon sock. There was definitely more sand incorporated with the sludge in the tubes wrapped with the 20 g/m^2 Cerex and 27 g/m^2 Reemay.

Both the Reemay and the Cerex had suffered some abrasion damage prior to installation. Attempts were made to patch the damaged envelope material as the tubing was rolled out immediately before installation, but all holes may not have been patched. In the places where the tubing was dug up, the Reemay and Cerex were both found to be fragile, whereas the nylon sock and Typar had good strength and were bridging over the corrugation valleys.

Drainage Rates

The outflow from each lateral was measured at various times following periods of rainfall and irrigation. The water table height at mid-spacing between drain laterals was also measured concurrently with the measurement of the drain outflow. It was recognized that to avoid interactions between drains it would be desirable to have three adjacent drains with each envelope type in each replicate. Discharge measurements could then be made on only the centre lateral of each group of three. The other two laterals in each group would merely provide buffer space between the drain envelope treatments. Unfortunately, we did not have the money to pay for three times as much drainage installation. Also the farmer did not wish to interrupt a larger area from his planned cropping pattern. Since it was not expected that the differences in drain discharges or rates of water table drop would be large, it was decided to install the drains with the randomization as shown in Fig. 1. The outflow from each drain was considered to be due to the average of the heads at mid-spacing between the two adjacent drains. A typical sample graph of the 1977 outflow measurements vs the average of the water table heights at mid-spacing between two adjacent drains is given in Fig. 4.

From Fig. 4 one can see that drainage rates exceed 15 mm/day when the water table is more than 770 mm above the drain bottom invert. Fifteen mm/day would be considered a very good drainage rate for sandy soil for potato or vegetable growing in this region. Considering the head required to cause a flow of 15 mm per day, it can be seen from Fig. 4 that a head of 1205 mm is needed for the tubing with small slots and no envelope while a head of 750 mm is needed for tubing with the Typar wrapping or 920 mm for the tubing with the nylon sock.

Looking at Fig. 4 another way it can be seen that for a head of 750 mm the flow from the tubing with the small slots is only 5.5 mm/day, from the tubing wrapped with Typar is 15 mm/day, and from the tubing with nylon sock is 9 mm/day.

Generally, Fig. 4 shows that the discharge rate from tubes with envelope material is greater than from those without envelope material for the same average water table level at mid-spacing. The tubes with envelope material appear to have a reduced entry resistance due to the increased inflow area at the drain tube surface.

Water Table Drop

Fig. 5 gives a picture of typical measurements of water table position in the 3 days following the end of irrigation or rainfall. Similar

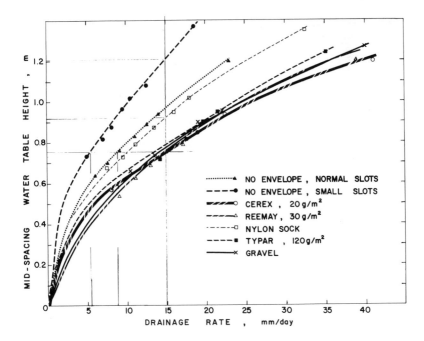

Fig. 4. Drainage Rate Versus Water Table Level at Mid-Spacing
Between Drains Along Piezometer Line B, Replicate 3
After Irrigation, August 24-28, 1977

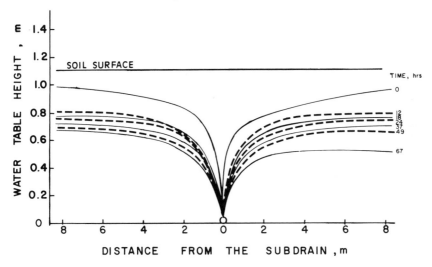

Fig. 5. Observed Water Table Heights Above Bottom Invert
of Lateral Drain 18 Piezometer Line B Following
A Soil Saturating Irrigation in August 1977

measurements were made on all 21 lateral spacings for 2 piezometer lines, 2
irrigation periods and one rainfall period in 1977. From Fig. 5 the water
table at mid-spacing between drains is seen to fall from 990 mm to 700 mm
above the bottom invert of the drain in 24 hours after the end of irrigation. This would provide very good drainage for most crops. The water

table drops less rapidly as it approaches drain depth, as can be expected due to the reduced head forcing flow.

Entry Resistance

From Fig. 5 it can be seen that about 65 percent of the head loss occurs within one meter of the drain tubes.

It is normally expected that about 50 percent of the head loss would occur in the first 20 percent of the spacing away from the drains due to the higher velocity of the water as it approaches the drain lines.

The very steep slope of the water table near the drains could be due to a decrease in the hydraulic conductivity of the soil in the field with depth, the development of a flow-impeding layer around the drain tubes, a blocking of part of the drain tube or envelope material entry area with microbial sludge growth, or differences in the entry resistance characteristics of the envelope materials used.

Observations in 1979 and 1980

Observations of drain discharge and water table height were made following some rainfalls in the autumn of 1979. The sediment and ochre in the pipes was observed in the summers of 1979 and 1980.

Sediment and Ochre in the Drain Tubes: The tubes with normal slots and with small slots but no envelope material were essentially blocked with the fine sandy soil by the summer of 1980. The tubes with an envelope material contained almost no sediment but they all contained significant amounts of ochre. The soil contained large amounts of ferous iron which favoured ochre formation. Much ochreous material came out of the drainage tubes during the periods of high water tables in the spring and autumn of 1978, 1979 and 1980.

Drainage Rates: Drainage rates for all treatments have decreased significantly over time. Fig. 6 shows drainage rates and center line water table heights measured in the 1979 autumn. From the measurements shown in these graphs it is evident that for a given water table height the drainage rates are only about one half of the values which occurred in the 1977 measurements.

The reduction in drainage rates for the tubes without envelopes is due to the tubes filling up with sandy soil, as well as a build up of ochreous material.

The difference in drainage rates with time for the tubes with envelope materials appears to be primarily due to the build-up of ochreous growths around the envelope.

Measurements about the ochre problem in this field were presented by Gameda, Jutras and Broughton (1981).

CONCLUSIONS

1. Of the envelope materials used in this investigation the 119 g/m^2 Typar (spun bonded polypropylene), the knitted nylon sock and the gravel performed satisfactorily at allowing water to enter the drain tubes but preventing the fine sand from entering the drain tubes.

2. Drain tubes without envelopes filled with fine sand. The small slots 0.7 mm wide in one kind of drain tubing were not small enough to keep the sand from flowing into the drain tube.

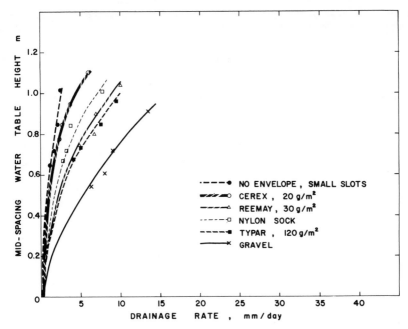

Fig. 6. Drainage Rate vs Head at Mid-Spacing
Along Line B, in Replicate 3, September 1979

3. The Reemay and Cerex envelope materials suffered much abrasion damage in transport and handling before installation, and were very flimsy when dug up in the field.

4. Ochre deposits grew inside and outside the drain tubes in this field. The ochre most probably was responsible for the decrease in drainage rates for a given water table height which occurred between 1977 and 1979.

5. To determine whether the drainage rates could be expected to decrease over the years for different envelope materials in soils where ochre formation is not a problem, a separate experimental installation is needed in a sandy soil with a very low iron content.

ACKNOWLEDGEMENTS

The authors gratefully acknowledge the financial assistance of the following companies which donated drainage materials:

    Sherbrooke Pipe and Drainage Limited
    Plastidrain Limiteé
    Modern Drainage Systems
    Montréal Terra-Cotta Limitée
    Daymond Limited
    Big "O" Plastics Québec Limited
    Drainbec Limitée

The careful work of the construction crew of Drainage Alseau Limitée who installed the experimental drains with a Speicher Trencher is recognized.

The work of R. Dehaan, C. Madramootoo, R. Langlois, R. Asselin, S. Ami, C. Damant, B. English, R. Munro, D. Valois, C. Nolet, R. Beaulieu, G. Bolduc and others who have participated in some of the installations, irrigation, measurements and observations is acknowledged.

REFERENCES

1. Broughton, R.S., B. English, C. Damant, S. Ami, E. McKyes and J. Brasseur, 1976. Tests of filter materials for plastic tubes. Proc. Third National Drainage Symposium, ASAE, St. Joseph, MI, U.S.A.

2. Gameda, S., P.J. Jutras and R.S. Broughton, 1981. Ochre deposition in a Québec fine sandy soil. ASAE Paper No. 81-2543, ASAE, St. Joseph, MI, U.S.A.

STRUCTURAL STABILITY OF SOIL AND THE NEED

FOR DRAINAGE ENVELOPES

W. Dierickx

Although envelope materials surrounding subsurface drain pipes considerably reduce entrance resistance (Nieuwenhuis and Wesseling 1979, Dierickx 1980), they are mainly applied to prevent significant soil particle movement into the drains. Therefore, envelope materials are used in unstable sandy soils while stable structured cohesive clay soils do not require envelopes. For intermediate soils, the need of envelopes depends on soil cohesion and structural stability.

Distinction is made between fine and coarse textured envelopes. Fine textured envelopes can be thin or thick while coarse textured ones are usually more voluminous. For cohesionless sandy soils, Ogink (1975) already proved that envelope pores may be larger than soil particles due to the arch building effect and the apparent cohesion of irregularly shaped particles. The criteria established for sandy soils certainly do not hold for weakly aggregated soils susceptible to erosion. Therefore, a weakly aggregated soil was used to determine the effect of various textured envelopes on the risk of clogging and on soil particle movement.

Aggregate stability does not only depend on soil composition; it is also influenced by the moisture content and the forces exerted on the aggregate. Flowing water exerts a force in the flow direction that is proportional to the hydraulic gradient. High gradients can deteriorate aggregates, causing internal blockage of pores and reducing the hydraulic conductivity. Willardson (1979) defines the gradient at which these phenomena occur as the critical hydraulic gradient and relates it to the D_{60} size of the soil. The performance of envelope materials has been investigated taking into account the effect of initial moisture content on aggregate stability and of hydraulic gradient on aggregate deterioration for a particular soil.

EXPERIMENTAL PROCEDURE

In the vicinity of subsurface drain pipes water flows radially from all directions. This flow approach has been simulated in sand tank models but seems rather difficult to achieve for structurally unstable soils. To overcome this problem the flow approach can be simplified to a one-dimensional flow experiment with a restricted amount of soil on the condition that the most unfavorable situation is simulated.

Water flow exerts a pressure in the flow direction. Downward flow acts in the direction of the gravitational force and drags the soil particles downwards tending to stabilize the soil. The lifting action of upward flow, on the contrary, promotes an unstable situation as a result of two antagonizing forces. Hence the upward flow experiment simulates the most unfavorable situation. To get evidence of this phenomenon both downward and upward flow experiments were performed (Fig. 1).

The plexiglass cylinders had a height of 300 mm and a diameter of 100 mm. In

The author is: Dr. Eng. W. DIERICKX, Senior Research Officer, National Institute of Agricultural Engineering, Merelbeke, Belgium.

the downward flow case (Fig. 1a) a screen supports the drain, envelope material and soil sample. The drain area was 78.5 cm² with an inlet area of 1.3 cm² distributed over 25 perforations which corresponds with a perforation area of 25.6 cm²/m of drain length. For the upward flow (Fig. 1b) the plexiglass cylinders can be considered as turned upside down after installation of a supporting screen protected by a fine filter material. Piezometric tubes were connected every 25 mm and the whole system was constructed in such a way that one tube was situated at the interface of soil and envelope material.

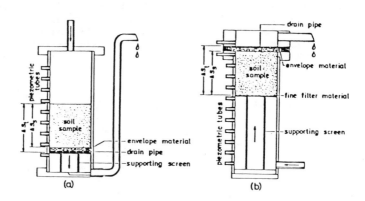

Fig. 1   Schematic Presentation of the Apparatuses for Investigating Downward (a) and Upward (b) Flow

The experiments normally ran for 9 days. However, some experiments had a shorter duration, due to the maximum obtainable hydraulic gradient which was reached quickly for low permeable soil samples to get a measurable discharge, or due to the maximum obtainable water supply which was the limiting factor for highly pervious soil samples.

The small amount of soil required allows one to condition soil samples for replications under identical circumstances. A limited height of the soil sample allows the application of high hydraulic gradients necessary to cover the range of hydraulic gradients which can occur near the drain due to the convergency of the streamlines (Dierickx 1980).

SOIL PREPARATION

To compare the results of various experiments with the same soil, the soil has to be prepared in the same way. The method usually applied at the laboratory of the National Institute of Agricultural Engineering consists of passing air-dried soil clods through a 5 mm square hole sieve. These aggregates were brought to a moisture content slightly higher than the desired one by spraying water with a paint gun while mixing. The moist aggregates were sieved again through 4.76, 3.36 and 2.00 mm square hole sieves. Aggregate samples were then reconstructed in such a way that natural aggregate distribution is approached. The soil samples were allowed to dry until the desired moisture content.

The soil used by Dierickx and Yüncüoğlu (1982) to investigate the factors affecting the performance of drainage envelope materials was a silt loam soil (U.S.D.A.-classification; Spangler, 1963) with the following particle size distribution: 16.5 % clay (0-2 µm); 61.4 % silt (2-50 µm); 22.1 % sand (> 50 µm) and an organic carbon content of 0.4 %. For this particular soil, samples were prepared using the following aggregate fraction: 40 % between 0 and 2.00 mm; 40 % between 2.00 and 3.36 mm and 20 % between 3.36 and 4.76 mm.

FACTORS INVESTIGATED

Besides flow direction, Dierickx and Yüncüoğlu (1982) used soil sample heights of 100, 150 and 200 mm to determine the effectiveness of envelope materials under various initial moisture conditions and increasing hydraulic gradient. Soil samples with initial moisture content of 3(air-dried), 5, 10, 15, 17.5 and 20 % by weight were used and from these, two unstable aggregate situations were taken to determine the effectiveness of envelope materials.

Both coarse and fine textured envelopes were used. The coarse textured envelopes were coconut fibres, synthetic fibres having a similar structure and a perforated sheet with polystryrene balls. The fine textured envelope consisted of needled non-woven synthetic fibres. The pore size distributions of these envelope materials are given in Fig. 2.

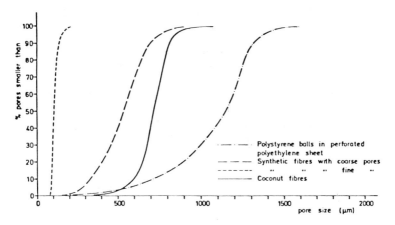

Fig. 2   Pore Size Distribution of the Envelope Materials
Used in Unloaded Situation

RESULTS OBTAINED

Before starting each flow experiment, the soil column was saturated slowly from below to remove the air from the soil sample.

Piezometric heads within the soil were very irregular and not suitable for analysis. Therefore only head losses over the soil column, Δh_s, and the overall head losses over the soil, the envelope and the flattened drain pipe, Δh_t, were used to compute average hydraulic conductivities k_s and k_t from discharge measurements.

The results obtained by Dierickx and Yüncüoğlu (1982) on flow direction, carried out on soil samples with initial moisture content of 3, 10, 15 and 20 % and protected by coconut fibres, confirm the above-mentioned statement that upward flow simulates the most unfavorable situation. The downward flow case results in higher k_s-values attributed to its stabilizing effect and more pronounced differences between k_s and k_t-values due to compaction of aggregates in contact with the envelope material. Therefore only upward flow was further considered.

Soil sample height was investigated on samples with an initial moisture content of 15 % and coconut fibres as envelope material. During the experiments the hydraulic gradient i_t being $\Delta h_t/\Delta s_t$ was increased. At the same gradient almost identical k_s-values (Fig. 3) and differences between k_s and k_t-values were obtained. A height of 100 mm was chosen for further experiments.

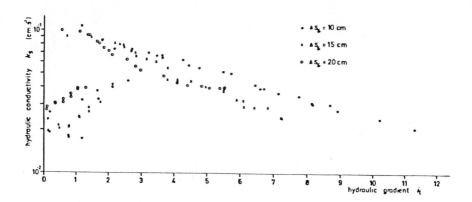

Fig. 3 Hydraulic Conductivity as a Function of the Hydraulic Gradient
for Three Soil Sample Heights

The initial moisture content exerts an important influence on aggregate stability and consequently on the hydraulic conductivity. Dry soil aggregates are
particulary vulnerable to deterioration upon wetting and low values of hydraulic conductivity result. The increase of hydraulic conductivity with increasing initial moisture content, shortly after starting the experiments, can
be seen in Fig. 4.

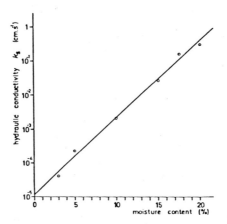

Fig. 4 The Hydraulic Conductivity as a Function
of the Initial Moisture Content of Aggregates

Because wetter aggregates become weaker, slightly lower values of hydraulic
conductivity were obtained at the end of the experiments for aggregates with
an initial moisture content of 20 % compared to those of 17.5 %. Up to moisture contents of 15 % differences between k_s and k_t-values were negligibly
small. Differences could be established clearly at moisture contents of 17.5
and 20 % due to some degradation of aggregates in contact with the envelope
material. This resulted in a somewhat lower overall hydraulic conductivity
k_t.

Because dry aggregates were already destroyed by wetting and moist aggregates
withstand the action of increasing hydraulic gradient, the hydraulic conducti-

vity of soil aggregates with initial moisture contents of 3, 5, 17.5 and 20 %
remains rather constant. Aggregates with moisture contents of 10 and 15 % are
found to be sensitive to deterioration with increasing hydraulic gradient
(Fig. 5), the more so as less aggregates were destroyed by wetting. The short
rise of hydraulic conductivity at low gradients is attributed to the removal
of fine soil particles from pores between aggregates. Dry aggregates which
were already destroyed by wetting do not give this increase at low hydraulic
gradients.

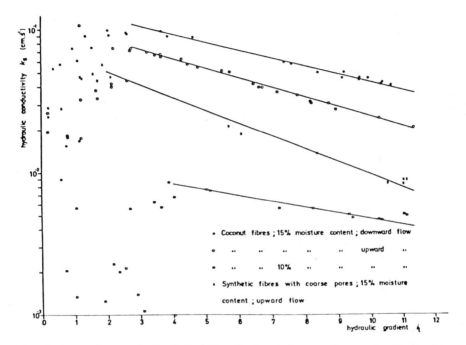

Fig. 5 Dependency of the Hydraulic Conductivity on Increasing Hydraulic
Gradient at Moisture Contents of 10 and 15 % (Dierickx and Yüncüoğlu, 1982)

From the above mentioned experiments it follows that soil conditions can greatly differ. The most unfavorable situation should be considered in determining
the effect of envelope materials on clogging and on soil particle invasion.
Therefore Dierickx and Yüncüoğlu (1982) carried out experiments on dry aggregates (initial moisture content of 3 %) sensitive to deterioration upon wetting and on aggregates with an initial moisture content of 15 % particularly
vulnerable to increasing hydraulic gradients.

The results of both soil conditions in combination with various envelope materials are presented in Figs. 6 and 7. Using an initially dry soil (3 %
initial moisture content), all envelopes act in the same way and do not exhibit differences between k_s and k_t (Fig. 6). Soil aggregates with an initial
moisture content of 15 % result in additional head losses and consequently in
a lower overall permeability for the fine textured envelope as well as for the
polystyrene balls in perforated sheet. The coarse textured coconut and synthetic fibres do not exhibit additional head losses (Fig. 7). The variations
in hydraulic conductivity on Fig. 7 illustrate the combined action of increasing hydraulic gradient and soil particle movement. A sudden rise of the hydraulic conductivity indicates a break-through of soil particles while a gradual
increase supposes a continuous movement of detached soil particles. The increase in hydraulic conductivity is followed by a decrease due to increasing
hydraulic gradients, the soil particle movement being continued.

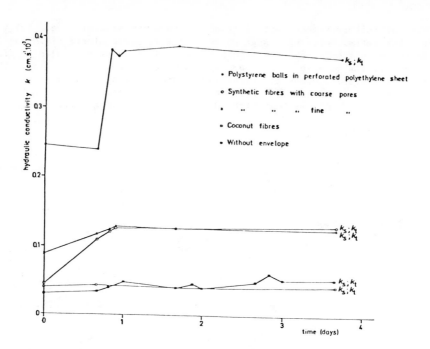

Fig. 6 Hydraulic Conductivity Values k_s and k_t at Initial Soil Moisture
Content of 3 % Using Various Envelope Materials (Dierickx and Yüncüoğlu, 1982)

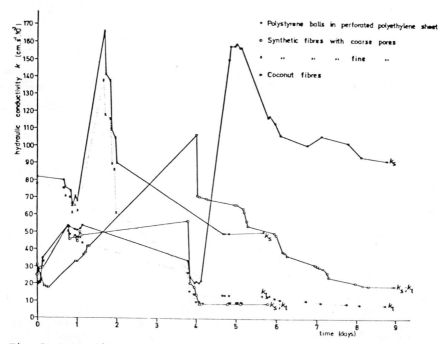

Fig. 7 Hydraulic Conductivity Values k_s and k_t at Initial Soil Moisture
Content of 15 % Using Various Envelope Materials (Dierickx and Yüncüoğlu, 1982)

Soil particle movement depends on the hydraulic gradient applied, and once started it increased gradually with increasing hydraulic gradient. At a given hydraulic gradient soil particle movement proceeded until equilibrium was obtained, time having no influence on further movement. Soil deposits in the drain and in the envelope material are given in Table 1. At initial moisture content of 3 % soil deposits were found in the drain and in a limited way in the envelope materials. Without envelope material, soil particle movement was restricted also but a complete piping with increasing hydraulic gradient occurred in the soil with an initial moisture content of 15 %. Polystyrene balls in a perforated sheet resulted in the highest amount of deposits in the envelope which reduces, together with the impervious part of the sheet, the proper functioning of the envelope. The soil particle accumulation outside the fine textured envelope creates a less permeable layer which is responsible for the additional head loss. Coarse textured envelopes allow a limited soil particle movement without clogging.

Table 1. Weight of Soil Deposits in the Drain and in the Envelope Material
(Dierickx and Yüncüoğlu, 1982)

| Envelope material | Weight of soil deposits (g) for soil initially dried to a moisture content of | | | |
| | 3 % | | 15 % | |
	In the drain	In the envelope	In the drain	In the envelope
Without envelope	0.48	–	piping	–
Polystyrene balls in perforated polyethylene sheet	10.15	0.67	6.97	5.02
Coconut fibres	8.05	1.42	3.25	1.84
Synthetic fibres with fine pores	0.11	0.70	0.05	1.04
Synthetic fibres with coarse pores	7.85	0.50	5.17	1.90

CONCLUSION

The need for envelope materials and the kind of envelope to use have been based on subjective criteria. To solve the problem simplified laboratory investigations can be carried out taking into account all possible factors influencing the proper functioning of drainage materials such as flow direction, soil sample thickness, soil conditions and the hydraulic gradient. Applying a one-dimensional flow the most unfavorable flow direction seems to be the upward flow. Soil sample thickness does not have any effect. Soil conditions are of great importance. Because the initial moisture content determines aggregate stability, the hydraulic gradient is important also because high hydraulic gradients can destroy aggregates.

The method described seems to be very useful to check these various parameters in order to decide whether or not an envelope material is needed and the kind of envelope material to use.

REFERENCES

1. Dierickx, W. 1980. Electrolytic analogue study of the effects of openings and surrounds of various permeabilities on the performance of field drainage pipes. Communications of the National Institute of Agricultural Engineering, Merelbeke, Belgium: nr 77, 238 p.

2. Dierickx, W. and H. Yüncüoğlu. 1982. Factors affecting the performance of drainage envelope materials in structurally unstable soils. Agric. Water Managem. 5 (In press).

3. Nieuwenhuis, G.J.A. ,and J. Wesseling. 1979. Effect of perforation and filter material on entrance resistance and effective diameter of plastic drain pipes. Agric. Water Managem. 2: 1-9.

4. Ogink, H.J.M. 1975. Investigations on the hydraulic characteristics of synthetic fabrics. Publ. 146, Delft Hydraulics Laboratory, 17 p.

5. Spangler, M.G. 1963. Soil Engineering. Second Printing. International Textbook Company, Pennsylvania, 483 p.

6. Willardson, L.S. 1979. Synthetic drain envelope materials. In: Proceedings of the International Drainage Workshop edited by J. Wesseling. ILRI, Wageningen: paper 2.02: 297-305.

ULTRAVIOLET STANDARD FOR CORRUGATED PLASTIC TUBING

G. O. Schwab   E. D. Desmond   W. J. Ochs
Fellow ASAE   Member ASAE   Member ASAE

ABSTRACT

A parallel plate impact test was developed to determine brittleness of 102-mm (4-inch) diameter corrugated plastic tubing when exposed to ultraviolet (UV) radiation. Artificial exposure was obtained with an Atlas Weather-Ometer Ci 35 using a xenon-arc lamp. Proposed exposure time was 200 hours which was selected to simulate two years or more of outdoor exposure. Samples from ten manufacturers were tested to evaluate proposed procedures. All samples passed the artificial UV exposure test, but three failed the parallel plate impact test due to a manufacturing problem at the mold seam and not due to embrittlement. Outdoor test results are available for only one-year of exposure.

INTRODUCTION

For marketing purposes and for reducing the loss of strength associated with high temperatures, corrugated plastic tubing (CPT) is made in white, red, yellow, gray, or other light colors rather than black. Light colored tubing may be less weather resistant, but black tubing also may be similarly affected if made with poor quality or improperly dispersed carbon black in the resin. Although tubing does not weather in the soil, it must retain its strength during the normal storage period in the distribution chain up to the time of installation. For these reasons industry, designers, and users generally are agreed that a standard for ultraviolet resistance is needed. This study was undertaken to secure laboratory and outdoor exposure test data from which such a standard could be developed. Such a standard is likely to become a part of ASTM F405 and/or a design standard of the U.S. Soil Conservation Service.

Salaries and research support provided by State and Federal Funds appropriated to the Ohio Agricultural Research and Development Center, The Ohio State University. Approved as Journal Article No. 146-82. Project SS-308.

The authors are: G. O. Schwab, Professor, E. D. Desmond, Technical Assistant, Agricultural Engineering Dept., Ohio Agricultural Research and Development Center and Ohio State University, Columbus; W. J. Ochs, Water Management Engineer, U.S. Soil Conservation Service, Washington, D.C.

Acknowledgment: The authors wish to express their appreciation to the Corrugated Plastic Tubing Assn. for financial support, to Hancor, Inc. for loan of the Weather-Ometer, and to other tubing manufacturers for providing tubing for testing.

REVIEW OF LITERATURE

General influences of ultraviolet radiation on plastics are well known. Furno (1980) summarized the effects of outdoor weathering of polyethylene (PE) for the Corrugated Plastic Tubing Association. He reported that actinic degradation in PE appeared as a reduction in molecular weight due to the breaking of carbon-to-carbon bonds. As molecular weight dropped, the plastic became more brittle. Factors that inhibit UV degradation were increased density, thicker walled materials, UV stabilizers, pigments, and antioxidants. By optical microscopy analysis, Montsinger (1977) found that lower extrusion rates improved the pigment distribution within the resin.

Weather-Ometer-outdoor exposure correlations are so material specific that only very loose correlations are available. Martinovich and Hill (1967) in a report on polyolefins for the Phillips Petroleum Co., found a correlation based on tensile strength which was recommended as a rough guide. They found that 1000 hours in the Weather-Ometer was equivalent to one year outdoor exposure in Arizona and more than 3-1/2 years in Ohio.

A Bell Laboratories study (Howard and Gilroy 1968) of low density polyethylenes with 2% carbon black obtained an expression relating Weather-Ometer exposure to outdoor exposure in New Jersey. The expression is $t_a = B + t_u^k$, where t_a and t_u are time of artifical and natural exposure in hours and years, respectively. B and k are constants specific to the weathering apparatus, location, and material. Criterion used to obtain B and k was the time required for samples to become brittle at a temperature of $-25°C$. Their values were 150 and 2.4, respectively.

Montsinger (1977) developed a brittleness test for high density PE in which the tubing is deflected between parallel plates to 50% of its inside diameter. The proposed ASTM standard for CPT brittleness is being developed to replace the environmental stress cracking specifications in ASTM(1977)F405-77a.

PROCEDURE

Test procedures for this study were developed and supervised by an advisory board consisting of representatives from the tubing industry, a government agency, and university research and extension faculty. Exposure tests were conducted with a xenon-arc lamp in the laboratory as prescribed by ASTM(1976) D2565-76, using a commercial Atlas Ci 35 Weather-Ometer made by the Atlas Devices Co. in Chicago. The lamp produced an irradiance level of $0.55 \text{ W/m}^2/\text{nm}$ at a wavelength of 340 nm. As shown in Fig. 1, the intensity of the lamp is similar to sunlight. The Weather-Ometer held twenty-one 102-mm (4-inch) diameter, 150-mm (6-inch) length samples on special racks. Tests were conducted without water spray for durations ranging from 100 to 400 hours.

Outdoor exposure tests were made at Columbus, Ohio, and at Phoenix, Arizona (DSET Laboratory). At Columbus exposure times were 12 and 24 months, and in Arizona the times were 6, 12, and 18 months. Exposure was made on stationary racks 45° from the horizontal facing south and as otherwise prescribed in ASTM (1975)D1435-75. Three tubing samples were tested for each exposure time.

Fifty tubing samples were furnished by tubing manufacturers for the tests. After artificial and outdoor exposure, the samples were placed in a refrigerator at $-10°C. \pm 0.5°C.$ for a minimum of 30 minutes. The samples were then removed from the refrigerator and within 15 seconds were impact tested using 4.84 kg-m (35 foot-pounds) of energy (21 pounds dropped from a height of 1.67 feet). The parallel plate impact tester is shown in Fig. 2. It has a quick release pin and nylon slides to reduce friction. During exposure the thinnest corrugator mold seam was nearest the light source, but for impact testing the line through the two mold seams was parallel to the plates on the

88

impact apparatus. On the lower plate a small groove was made in the plate to

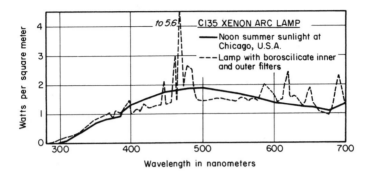

Fig. 1.   Energy spectrum of an xenon-arc lamp and noon summer sunlight
           at Chicago, Ill.   From Atlas Electric Devices Co. (1981).

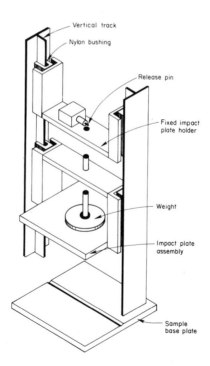

Fig. 2.   Parallel-plate impact test apparatus.

keep the sample from rolling on the plate. Failure from the parallel plate impact test is any crack or split more than 10 mm in length, except when crossing a corrugation, then failure is a crack or split extending from root to root or from crown to crown. The failure and retest requirements are shown in Table 1.

Table 1. Failure and Retest Requirements

Samples to be tested	Number passing	Number failing	Lot test results
7 (original)	7	0	Pass
7 (original)	6	1	Pass
7 (original)	5	2	Retest
7 (original)	4 (or less)	3 (or more)	Fail (no retest)
7 (retest)	7	0	Pass
7 (retest)	6	1	Pass
7 (retest)	5 (or less)	2 (or more)	Fail (no retest)

ARTIFICIAL UV TESTS

Initially, a 100-hour Weather-Ometer exposure duration was selected, but it was found that even nonpigmented polyethylene samples passed the impact test after exposure. Nonpigmented tubing was then irradiated for various durations to find the time for brittleness failure. Brittleness was detected at about 175 hours. The Weather-Ometer exposure duration was then increased to 200 hours, which was the time selected for the proposed outdoor exposure standard. White, black, yellow, and red samples were likewise tested for 200, 225, 300, and 400 hours. None of these samples showed any significant effect even at 400 hours of exposure.

Using the 200-hour exposure length, ten manufacturer's lots were tested. No lot showed any change due to UV exposure. However, in lots 4, 7, and 10 (Table 2) both exposed and unexposed samples failed. These failures were not due to ultraviolet exposure, but were attributed to a manufacturing problem at the mold seam where the failures occurred.

Table 2. Parallel Plate Impact Tests for Brittleness

Mfr. Lot No.	Color	Unex- posed	Artificial UV exposure (200 hr.)	Outdoor Exposure (1 year) Ohio	Ariz.
1	black	P	P	P	-
9	black	P	P	-	-
10	black	F	F	-	-
7	black	F	F	-	-
5	yellow	P	P	P	P
6	white	P	P	P	-
2	white	P	P	P	P
3	white	P	P	P	F
4	red	F	F	F	F
8	red	P	P	P	-

Note: Conditioned at -10°C. and impacted at 4.84 kg-m (35 ft-lbs).
P passed; F failed; - no test results.

In 1973-75 preliminary pipe stiffness tests were made with ten lots of 102-mm (4-inch) diameter, 150-mm (6-inch) length samples of six samples each. Half of the samples were exposed to full sunlight at Columbus, Ohio, and half were stored indoors (unexposed samples). As indicated in Table 3, black PE tubing showed an average reduction in pipe stiffness of 18.5% and 19.5% for 5% and 10% deflections, respectively, during the two-year period. Reduction in stiffness was nearly the same for the unexposed and the exposed samples. For the red samples the average reduction was about 11%, but the polyvinyl chloride (PVC) tubing increased an average of about 5% in stiffness for a one year period. These tests indicate that sunlight exposure did not affect pipe stiffness, however, parallel plate stiffness is not a test for brittleness. For this reason the parallel plate impact test was developed.

Table 3. Effect of Sunlight Exposure on Pipe Stiffness

| Tubing material | Color | *No. sample lots | Exposure time (years) | Percent stiffness change in two years | | | |
| | | | | 5% deflection | | 10% deflection | |
				Unexposed	Exposed	Unexposed	Exposed
PE	Black	7	2	-18.7	-18.3	-19.8	-19.2
PE	Red	1	2	-11.1	-11.5	-10.5	-12.0
PVC	White	2	1	7.9	5.2	2.9	4.0

Note: As per ASTM(1977)F405-77a, Parallel Plate Stiffness.
* Each lot consisted of six 102-mm (4-inch) diameter, 150-mm (6-inch) length test samples.

Outdoor exposure tests were made from the same lots that were artificially exposed in the Weather-Ometer. One of the objectives of the artificial exposure tests was to predict the time for the equivalent of two years of outdoor exposure. After one year of outdoor exposure, only lot 4 showed a significant increase in brittleness (Table 2). All lots were exposed outdoors in Ohio, but only selected lots were exposed in Arizona. Test results are not yet complete.

DISCUSSION

In order to conserve time for testing, the artificial UV exposure test was selected, but to be meaningful it must be related to outdoor exposure time. This study as well as previous investigations has shown that a simple relationship for all materials is not likely. The 200-hour test was selected using nonpigmented samples from two manufacturers. Although the resin in the lots was similar, thickness and perforations were not the same, which produced different results. Both lots passed the 100-hour test indicating this duration was too short. A minimum exposure time is desired to reduce operating costs, especially when tests are conducted by commercial laboratories. Energy and maintenance costs are rather high. It was necessary to select a test period (200 hours) early in the study in order to meet the approval date for nonblack tubing set by the U. S. Soil Conservation Service (SCS) for in-service use. The 200-hour test results will be the basis for SCS approval. Hopefully, the two-year outdoor exposure tests will validate the artificial test time. An important criterion for whatever standard is developed is that good tubing will pass and bad tubing will not. These test procedures are similar to those being considered for the revision of ASTM (1977)F405-77a (Project 3203, revised ballot A, March 1981), and hopefully a final standard will be cooperatively developed.

The number and size of samples for testing was limited by the Weather-Ometer capacity. A sample size of seven permitted three lots to be exposed at the same time. Special racks in the machine were made to hold twenty-one 102 mm (4-inch) diameter, 150 mm (6-inch) lengths of standard tubing. Since other sizes will not fit in the machine, a manufacturer will be required to produce 102-mm (4-inch) tubing from the same resin used in making the size to be approved.

Test data on exposed and unexposed CPT was included to show that the pipe stiffness test (ASTM(1977)F405-77a) is not a good test for brittleness. It did reveal a sizeable decrease in stiffness for PE tubing over a two-year period. Further investigations are needed to determine if such a reduction continues for more than two years and to update the data for current manufactured PE and PVC tubing as the tubing tested was made 10 years ago.

REFERENCES

1. ASTM. 1975. Standard recommended practice for outdoor weathering of plastics, D1435-75. Philadelphia, Pa. The Society.

2. ASTM. 1976. Standard recommended practice for operating xenon arc-type (water-cooled) light and water exposure apparatus for exposure of plastics D2565-76. Philadelphia, Pa. The Society.

3. ASTM. 1977. Standard specification for corrugated polyethylene (PE) tubing and fittings, F405-77a. Philadelphia, Pa. The Society.

4. Atlas Electric Devices Company. 1981. Ci 35 Weather-Ometer instruction booklet. Chicago, Ill.

5. Furno, F. J. 1980. Outdoor weathering of polyethylene. Prepared for Corrugated Plastic Tubing Association. Carmel, Ind. Litho.

6. Howard, J. B. and H. M. Gilroy. 1968. Natural and artificial weathering of polyethylene plastics. Polymer Engineering and Science, July, 1969, Vol. 9, No. 4, 286-294.

7. Martinovich, R. J. and G. R. Hill. 1967. Practical approach to the study of polyolefin weatherability. Applied Polymer Symposia, No. 4, 141-154.

8. Montsinger, L. V. 1977. Influence of processing conditions and resin characteristics on polyethylene corrugated tubing. Proceedings ASAE Third National Drainage Symposium, pp. 75-79. Chicago, Ill. Dec. 1976.

CAUSES OF PREMATURE FAILURE

AND PROPER INSTALLATION OF LARGE (25 cm to 75 cm)

SUBSURFACE DRAINAGE MAINS

Cy Schwieterman*
Member ASAE

As a subsurface drainage contractor doing installation in both Ohio and Indiana, it has always been a concern of mine, "Why are large tile mains nearly always broken when we want to connect a submain to them?" When we uncover the large main tile, we find the following: Of the mains installed 40 years or longer, over 95% are quartered or broken into 4 pieces. After the tiles have been broken for a number of years, the top pieces fall on the bottom of the tile and are sometimes moved down stream by the water flow. Occasionally these pieces form a block in the entire main. Landowners do have these large mains repaired by replacing only those that are completely broken, (Fig. 1). We do not recommend this practice as the main is deteriorated beyond repair at which time the complete main should be replaced. As you may know, in large areas of both Ohio and Indiana, and I'm sure in other sections of this country, large mains are an absolute necessity for subsurface drainage as Ohio has approximately 18,098 kilometers of mains in 59 counties.

Fig. 1 Broken Main

---

*Subsurface drainage contractor in West Central Ohio and East Central Indiana since 1946.

As Fig. 1 shows a broken down, collapsed and worn out main (very often before its alloted time) is not an asset to a farmer's subsurface drainage system. It is also a definite safety hazard as people, machinery, and animals do fall into these large holes thereby causing injury to people and breakage to equipment. Why this premature failure? As I look back, I can visualize why failure may have occurred earlier than expected.

1. No outlet pipes at open ditch, or headwalls that were not of
   sufficient size and strength.

Fig. 2 Poor Headwall

2. Improper installation. One of the reasons of collapse, due to
   improper installation, is that the ditch was not of proper width
   so the backfill did not bed properly in the void on the bottom
   half of the tile and/or the ditch bottom did not have the proper
   contour which is .5 of the outside diameter of the installed tile.

3. Poor outlets and poor maintenance on outlets and open ditches. A
   submerged outlet creates undue pressure on a system, thereby causing
   either suck or blow holes.

4. Poor quality tile and tile of insufficient strength. The ASTM C-4
   specification was adopted in the early 1900's. It was in the 1950's
   that a testing quality procedure was first set up in Ohio and Indiana.
   But a study of county petitioned mains showed that the majority of
   these large mains were installed in the years 1890-1915. At that
   time, large tile were not designed for the ever increasing load that
   farmers are and will be using in the future, as today's equipment is
   considerably heavier than 80 years ago.

5. Connection usually not of a manufactured type or, if hand made, poorly
   connected and mortared.

6. No provision for surface drainage.

7. No provisions for annual inspection and the necessary maintenance.

8. Large joint spacing.

9. Very poorly built and improperly placed inlets and receivers.

Now, after many years of experience, let us look at what we are still doing incorrectly.

1. Outlet pipes are not long enough but quality of recently installed headwalls and stone drops is very good.

2. Not wide enough ditch. Improper or no contour on bottom of ditch. Backhoe installation—ROUND tile on a FLAT ditch bottom without proper bedding. In bedding these large mains, we still find that the County Engineers (Professional Engineers) allow installation with backhoes with improper or no bedding. Even some Soil Conservation Service personnel are guilty of allowing this poor installation procedure.

3. Small amount of specifications and lack of their enforcement. In fact, some county engineer's specifications are almost non-existent.

4. Standard strength tile are still being installed.

5. Length of tile too short.

6. Gradual turns instead of elbows.

7. Still permitting large joint spacings or gaps.

8. Insufficient inspection during construction.

Because the installation and replacement of large tile mains have escalated in the last five years, I, as a contractor, saw the need for better installation, but still only about 15% of the jobs that we look at have proper written specifications for installation. None of the jobs we installed had the correct kind and amount of on-the-job inspection.

After finding considerable amounts of little or no inspection and improper installation and in a few cases poor designing, I thought that we should begin an educational program to correct these situations. So I made a list of the following 22 reasons for failure and the ways to prevent these failures and prolong the life of these large mains through proper planning and good installation practices.

1. Headwall, drop inlet, or stone structure of wrong size or improper construction.

   Develop a drainage plan for the watershed giving consideration to surface runoff, subsurface flow, inlets and receivers. The plan should meet either or all of the following:

   A. SCS Specifications
   B. State Highway Specifications
   C. Engineering Handbook Specifications
   D. State Drainage Guides
   E. ASTM Specifications

2. Outlet Pipe

   A. Type of material

      Aluminum or PVC pipe has a longer life. Use the next heavier gauge
      than is required. If steel pipe is desired, use bituminous pipe
      with a thicker layer flow line.

   B. Insufficient length

      Minimum length--12.19 meters

   C. Improper Size

      Use next size larger than tile, if needed to meet the required
      drainage coefficient.

   D. Does not have an anti-seep collar

      Anti-seep collars should be installed according to specifications
      listed in No. 1.

   E. Animal Guards

      Animal guard should be of the outside swing open type which helps
      prevent clogging and does not disturb the flow, and should be of
      the same type material as outlet pipe.

Fig. 3 Outlet Pipe and Animal Guard

3. Tile of wrong size for acres drained. Smaller than needed creates too
   much pressure. Larger than needed is more costly.

   Follow specifications listed in No. 1.

4. Tile of poor quality or insufficient strength results in premature
   breakdown of tile.

Due to increasing heavier machinery, EXTRA STRENGTH HEAVY DUTY or REINFORCED TILE should be used. All tile are to be tested yearly and should meet ASTM and/or state highway specifications. Manufacturer or bidder must present test data with bid. In organic soil, clay tile or plastic tubing should be used. Extreme care should be taken after installation so heavy machinery (bulldozers, earth movers, etc.) does not cross new tile line.

5. Tile installed with not enough cover resulting in premature breakage and possible frost damage.

   Following SCS specifications. If cover is insufficient, fill in low areas or,if possible or practical, install tile on higher ground. If the preceding alternatives are not used, use extra strength or reinforced pipe.

6. Tile installed with improper grade, resulting in decreased capacity or insufficient cover.

   Follow specifications as listed in No. 1 for proper grade and capacity. Inspection of grade should be at the time of installation. A Laser equipped machine is desirable, but with all machines the quality of work done depends on the knowledge and skill of the operator.

7. Tile installed with too large joint spacing resulting in suck or blow holes.

   If tile are laid with openings or gaps that are too large, cover the openings with a permanent type material such as coal tar pitch treated roofing paper, fiberglass sheet or mat, or plastic sheet (15cm-20cm wide) around the complete outside diameter of the tile. This prevents suck or blow holes that allow soil to enter tile. Pushing the tile together while installing is usually the answer. (See Fig. 4)

Fig. 4   Tile Ram

8. Length of individual tile not long enough. This results in more joint spacing and thereby increases the risk of suck or blow holes.

   The longer the tile the better.

20 cm, 25 cm, 30 cm.	-- Minimum length, 61 cm. Longer if available.
38 cm and above	-- Minimum length, 76 cm. 91 cm or 122 cm if available.

9. Tile settled off grade, resulting in decreased capacity and usually suck holes.

   In organic soil or very sandy soil conditions, use tongue and groove pipe or other interlocked pipe, including continuous length plastic. Use as long a length of pipe as possible.

10. Tile not installed on a round bottom ditch, or without a suitable contour in ditch bottom, or if installed with a backhoe, either insufficient or no granular backfill. This results in premature breakage because of placing a ROUND tile on a FLAT bottom ditch. (See Fig. 5)

Fig. 5 Broken Tile

Using a WHEEL MACHINE with a ROUND BOTTOM that exactly fits the outside diameter of the tile is ideal. If installing with a backhoe, the trench should be excavated 7.5 cm to 15 cm below grade and filled with granular material to grade, then tile laid and granular material filled to 180° of the outside diameter of the tile.

11. Installation of tile in a ditch that is of insufficient width results in no dirt along the bottom half of tile which reduces strength and allows flow outside of tile.

    Minimum width of ditch should be 30 cm wider (15 cm on each side) than the outside diameter of the tile.

Example for 38 cm tile:	Outside diameter	48 cm
	15 cm on each side	30 cm
	Width of ditch	78 cm

    Sloping ditches should begin with proper width 10 cm from bottom of ditch.

12. Poor backfilling, resulting in broken tile and/or no dirt along the bottom half of tile.

    Use of an auger backfiller is ideal. If backfilling is done with a blade, each push of the blade should not be over 122 cm wide.

13. Gradual turns instead of elbows, resulting in more space between tile and thereby causing more suck holes.

    Keep excavated trench straight. Elbows or manufactured fittings should be used for turns. If tile are sawed or ends of tile chipped to make an on-site elbow, the joints should be well sealed with concrete or mortar.

14. No waterway in the overall plan which can cause overloading of the main.

    Follow specifications listed in No. 1.

15. Main tile in waterway usually results in less than desirable cover.

    Keep tile to side of any surface drainage or waterway.

16. Tile does not have a free flowing outlet, which keeps tile either partially or full of water creating a back pressure and less capacity.

    The open ditch should be clean so that one foot of free-board is available.

17. Inlets and Receivers

    Inlets and receivers require more flow capacity in tile mains and should be installed only where absolutely necessary, such as, road ditches or in low or pond areas. Offset from main ditch if possible.

    A cast iron or steel grate should be attached on top to keep animals out and should be fastened securely so it cannot be stolen. Use sealed pipe and fittings for connecting to main tile. If the catch basin is in the farmed area of the field, mark with four posts around outside of structure 183 cm to 244 cm above ground level.

18. Breathers allow for tile to fill to capacity, and also serve as an inspection point.

Install breathers at property line 60 cm above ground level. Have
permanent breather on top fastened to prevent stealing.

19. Poor connections, causing suck or blow holes.

    All tile to be connected should have a manufactured fitting. If a
    hole is sawed or cut into the main, the connection should be well
    cemented. The first 450 cm of tile away from a main should not have
    any spacing or be solid sealed pipe.

20. Road and driveway crossings usually break down when plain drain tile
    are used.

    Use reinforced or vitrified pipe thru road. If plastic tubing is
    used, follow manufacturer's instructions for installing.

21. Maintenance - one ounce of prevention is worth a pound of cure.

    Ditch should be under a maintenance contract or checked yearly by a ditch
    commissioner or engineer for any needed maintenance, which should be
    performed as soon as practical.

22. Plastic mains - 25 cm and above.

    Follow manufacturer's specifications for installation. From past
    experience I recommend that plastic mains being installed should have
    particular care given to the following: proper blinding and bedding,
    allowing no stretching, positive locking of the couplers so they cannot
    pull apart, and auger type backfilling if a wheel machine is used.
    Contour of bottom of trench should conform to fit .5 diameter of pipe
    being installed. If solid plastic pipe (no drainage holes) is used in
    sand, the sand should be drained with a filter wrap on a smaller diameter
    pipe laid parallel to the proposed new main.

             IN SUMMARY - PROPER INSTALLATION

For many obvious reasons, the existing large subsurface drainage mains are
rapidly failing before their alloted time, so I recommend the following:

1. Develop a surface and subsurface plan for the watershed and properly
    design the complete large main system according to appropriate
    specifications.

2. I predict that in 25 years, farm equipment will be 15% to 25% heavier
    than today, so I recommend all of the following specifications be
    changed--width of trench should be wider, better bedding, longer length
    of tile, and tile with greater crushing strength. The blinding and
    bedding could be better implemented by allowing mechanical methods.
    Contour of bottom of trench should conform to fit .5 diameter of
    pipe being installed.

3. More rigid enforcement of specifications along with continuous
    inspection while construction is in progress. Remember, specifications
    are not worth the paper they are printed on if continuous on-site
    inspection is not done.

4. All new mains should have rigidly enforced rules for yearly inspection
    and the necessary maintenance performed as needed.

5. The use of plastic mains are increasing. Extreme care should be
    exercised in installation. Items of concern are the correct contour
    of the bottom of the trench, eliminate stretching as much as possible,
    use friable material for proper bedding before backfilling, auger backfill
    if an open trench is used for installation, let natural settling occur if
    at all possible. Do not allow large construction or farm equipment
    to cross backfilled ditch until natural settling occurs. Our experience
    is that it is extremely difficult to install plastic mains without
    stretching. To date we have not consistently installed large diameter
    tubing without stretching.

    From our experience, if solid plastic mains are to be installed in
    sand, I recommend that the sand be drained before installation.
    One way this can be accomplished is by installing smaller filter wrapped
    plastic tubing approximately 15 feet from the proposed line of the
    large main.

    Following manufacturer's specifications for the installation until
    the ASTM specifications are established.

Because large mains are one of the most permanent land improvement practices
that a landowner does (expected life is 20 to 80 years), more careful
inspection, installation, and maintenance should be exercised to give longer
life and better subsurface drainage resulting in more income to the landowners in the entire watershed.

DRAINAGE-PLOW "BOOT" DESIGN TO IMPROVE

INSTALLATION OF CORRUGATED PLASTIC TUBING*

J. L. Fouss   W. E. Altermatt
Member ASAE   Assoc. Member ASAE

The plow-in method of installation for corrugated plastic tubing gained
rapid acceptance and use world-wide during the 1970's. Under some soil
conditions, the quality of installation is poor with high-speed plows in
comparison with trench-installed drains, because positive blinding and
backfilling is not provided with the plow-type equipment. Newly developed
implement attachments for the drainage plow, namely, a hydraulic controlled
blinder/backfiller double disc assembly, and a revised design for the tube
feeder or "boot", can significantly improve installation quality for a wide
range of soils and installation conditions.

PLOW "BOOT" DESIGN

Most present-day standard drainage plow "boots" for installing 10-cm diameter
corrugated plastic tubing have an average width of 15-cm, and are approximately J-shaped in side-view. For drain installation at 2.5 to 3.5 meters
deep, this narrow plow-trench will often collapse together at the ground
surface as the "boot" passes, without significant soil backfill falling into
the void behind the "boot" -- essentially no top soil falls down through the
void to drain depth (top soil is commonly used to "blind" tubing installed
with trenching machines). In wet fine-textured soil conditions, the void at
drain depth may not be filled with soil at all, and the polyethylene plastic
tubing will float upward off the plow-trench bottom as water infiltrates into
the void; subsequent soil settlement sometimes surrounds the tubing at a
level above the design drain depth, thus creating grade-line irregularities
and potential sedimentation problems.

The revised design for the plow "boot" includes the following features: (a)
a minimum width of 23 cm for installation of standard 10-cm diameter corrugated plastic tubing; (b) a side-view shape like a 60°-30° right triangle,
with the rear surface (hypotenuse) at a 60° angle from horizontal; and (c) a
replaceable friction reducing surface (e.g., high molecular weight polyethylene plastic sheet) on the rear 60° - inclined, 23-cm wide, surface of
the "boot".

The authors are: J. L. Fouss, Agricultural Engineer, ARS-USDA, Baton Rouge,
LA, and W. E. Altermatt, Product Manager - Parts, American-Lincoln Div., a
Scott Fetzer Co., Bowling Green, OH (formerly, Chief Scientist and Asst.
Product Manager, respectively, with Hancor, Inc., Findlay, OH).

*The authors wish to acknowledge that the development reported herein
resulted from a project conducted while in the employ of Hancor, Inc.,
and is presented in the interest of potentially improving quality of
subsurface drain installations with plow-type equipment.

The width dimension for this revised "boot" is the same as that presently used with most plows for installing 15-cm diameter corrugated tubing. It is desirable to increase the plow blade width to about 26-28 cm in order that soil drag along the sides of the 23-cm wide "boot" is reduced to allow proper "floating" action during grade-control corrections [for details refer to Fouss (1982)]. The 23-cm wide "boot", with the 60°-inclined rear surface, maintains an opening, during forward travel, from the drain depth to the ground surface for downward passage of blinding and soil backfill; the downward movement is enhanced by the low friction, inclined rear surface on the "boot". Top soil and backfill soil are positively placed onto the top end of the "sliding-board" type rear surface of the "boot", during forward motion, with the double disc assembly described below. We did not conduct tests to determine the best angle for the inclined rear surface of the boot, but the 60° angle appeared to provide for uniform gravity flow of loose soil from the ground surface to drain depth. A smaller angle (e.g. 45°), which created a much longer "boot" base, tended to slow the soil flow down too much, and backfilling of the plow-trench was often not as complete as with the 60°-plane.

DISC BLINDER/BACKFILLER

The blinder/backfiller implement is simply a double disc arrangement on two independent control arms which are pivot mounted to the plow blade support frame. The disc assembly is not mounted onto the plow "boot" because the desired "floating" linkage action between most "boots" and plow blades would be hindered. The control arm for each gang-pair of disc coulters are independently operated with a hydraulic cylinder. An active pressure of about 2,000 kPa on the hydraulic cylinders ensures good cutting action, and a 2,400 kPa pressure release valve permits the discs to roll over a rock or very hard soil clod without damage.

Two styles of disc coulters are used on the blinder/backfiller assembly; for each gang-pair the larger disc coulter is positioned on the inside (closest to the soil void left by passage of the "boot"), and is a "conical" shaped disc of 60-cm diameter; the outside discs are the more common "dish" shape and of 45-cm diameter. Both disc types have notched edges to ensure more positive rotation and cutting while passing through soil where sod or crop residue is present. The spacing between the inner (larger) and outer (smaller) disc on each gang-pair is at least 30 cm; this protects against soil clogging between the discs on each gang-pair in sticky clay-type soil -- the difference in disc shapes also aids in this regard. [For more details see Fouss (1982).]

FIELD OPERATION

The proper adjustment and field operation of the Disc Blinder/Backfiller assembly in combination with the revised Plow "Boot" involves three primary factors: (1) the alignment or position of the disc gang-pairs with respect to each other; (2) the position of each gang-pair with respect to the inclined rear plane on the plow "boot"; and (3) the downward active hydraulic pressure and release pressure settings for the soil conditions.

These are accomplished by first adjusting the disc control arm lengths such that one gang-pair of discs trail the other so that soil moved into the "effective trench" left by passage of the boot is not "bridged" over at the top because the soil from opposing disc pairs merge. Second, the disc gang-pair running the closest to the rear of the boot should be positioned with respect to the inclined rear plane of the "boot", so that the soil movement from this disc pair falls directly onto the plane. This will ensure that <u>top soil</u> from the ground surface is placed on the sliding-board boot plane first, and this top soil will travel (slide) down the plane to cover the installed plastic conduit emerging from the "boot". The forward disc pair may need adjustment from time-to-time depending on the depth of drain installation and soil moisture conditions; the deeper the drain, the further forward the disc pair may need to run. The trailing gang-pair should be behind the first pair by 30 cm or more for most soil conditions. The lateral position of the disc gang-pairs should be adjusted for soil conditions to cause all soil movement directly into the "effective trench" void in the case of the leading pair, and into any remaining void for the trailing pair. The resulting ground surface is considerably more level and ready for subsequent cultivation than conventional plowed-in drains.

The active hydraulic pressure should be set at a level high enough to ensure good cutting of the discs, but not so high as to bury the disc to the axle, and thus cause poor backfilling and excessive draft. The release pressure can be set, as a rule-of-thumb, at about 400-500 kPa above the active pressure. The release pressure should also be at a value high enough to allow the disc to be lifted hydraulically into the transport (non-cutting) position. For safety purposes, the manual hydraulic control lever should be provided with a safety lock to ensure that it cannot be accidently moved from a given position (up or down). As an additional safety measure, to prevent the discs from being lowered (or "free-falling") too fast from the elevated transport position, flow regulator/check valves are placed in the appropriate hydraulic connections for each cylinder. Since the disc coulters are exposed, the equipment operator should ensure that other persons are away from the plow-boot area before moving the discs.

CONCLUSIONS

Our field experiences during 1981 with the disc blinder/backfiller and new plow "boot" design have led us to conclude:

(1)   The 23-cm wide plow "boot" for installation of 10-cm diameter corrugated tubing provides and maintains an opening, during forward travel, from the ground surface to the drain depth, through which top soil for blinding and backfilling can be effectively transported via a "sliding-board" mode on the 60° inclined rear surface of the "boot".

(2)   Operation of double disc gangs on each side and behind the plow "boot" effectively pulls loose topsoil onto the inclined soil transport plane of the "boot" to provide for positive blinding and backfilling of plowed-in drains; this procedure also provides a secondary benefit by leveling the heaved ground left by passage of the plow-"boot" -- farmers and contractors alike have been pleased with the performance of the disc assembly.

REFERENCES

1.   Fouss, J. L. 1982. Modernized Materials Handling and Installation Implements for Plow-type and Trencher Drainage Equipment. International Drainage Workshop, Dec. 6-11, 1982, 4-H Center, Washington, D.C.

TRENCHLESS DRAINAGE PIPE INSTALLATION AND ITS
IMPLICATIONS FOR SUBSEQUENT DRAIN PERFORMANCE

G. Spoor   R. K. Fry

The trenchless drainage installation technique is used widely in many areas
and frequently can offer significant cost advantages over other methods.
Some results from a number of field experiments, however, suggest there may be
circumstances where its use may impair subsequent drain performance, see
Eggelsman (1979), Naarding (1979) and Olesen (1979). All the problems high-
lighted in these experiments were associated with high .entry resistance for
water flow into the pipe and occurred in groundwater control situations in
fine textured soils. This suggests some form of soil compaction or smear in
the vicinity of the pipe. To date, the exact circumstances requiring caution
in the use of the technique have not been clearly defined. This paper
presents the results of a field and laboratory investigation aimed at identi-
fying the situations where problems may arise, their likely magnitude and the
ways by which they may be minimized or overcome. The investigation involved
the use of:

a. soil bin and field experiments to identify the main soil failure
   mechanisms generated by trenchless tines.
b. electrical analogue techniques to establish the effect of different
   degrees of compaction near a drain pipe on drain performance.
c. field tests to determine the actual soil disturbance caused by the major
   types of trenchless machines when working under different soil conditions.

INFLUENCE OF COMPACTION NEAR A DRAIN PIPE ON
DRAIN PERFORMANCE

The deleterious effects of compaction in the backfill material itself around
the pipe on drain performance have been clearly identified in Cavelaars
(1967). Electrical analogue investigations described in Fry and Spoor (1982)
have quantified similar effects resulting from impedance zones located below
and to the sides of the pipe beyond the backfill, at the drainage machine/
cavity interface. The analogue used assumed the impedance layers to be
impermeable and hence the results relate to an extreme case of compaction.
The results show that limited compaction occurring below the pipe only and
extending horizontally to a width of approximately two pipe diameters, has
little effect on the mid-drain water table position. Compaction beyond this
does increase water table height, although its effect is considerably less
than vertical compaction at the side of the pipe. Figure 1 shows the rapid
rise in mid-drain water table position with increasing height of side compaction. Any compaction or smear of this form produced during trenchless
installation could seriously reduce the effectiveness of the drainage scheme.
This form of compaction is less likely to occur in trenching installations
and hence could be the major cause for the poorer performance of the trenchless installations reported earlier.

The authors are: G. Spoor, Professor and R. K. Fry, Research Officer,
National College of Agricultural Engineering, Silsoe, Bedford, UK.

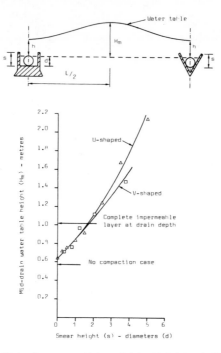

Fig. 1 Effect of Side Smear on Water Table Height at Mid-Drain Position

SOIL DISTURBANCE CAUSED BY TRENCHLESS MACHINES

Spoor and Fry (1982a) have shown that two types of soil disturbance can occur with trenchless tines. The actual type is dependent upon working depth, tine geometry and soil condition. Figure 2 shows in rear elevation the soil

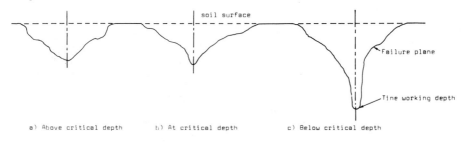

a) Above critical depth   b) At critical depth   c) Below critical depth

Fig. 2 Variation in Soil Disturbance Profile (rear elevation) with Increasing Tine Working Depth

failure profiles generated by a trenchless tine working over a range of depths. At shallow depths, the failure planes diverge rapidly from the tine base, the soil within the planes being heaved upwards and generally loosened. With increasing depth a "critical depth" is reached, defined in Godwin and Spoor (1977), below which the divergence decreases rapidly with only a narrow zone of soil being loosened between the failure planes. At working depths above critical depth, there is little soil disturbance beyond the failure planes and soil compaction is unlikely to occur in this area. The reworking of soil at the failure plane in low shear strength soils, may cause a little compaction at the interface. Below critical depth, significant lateral soil

disturbance can occur beyond the failure planes, which could result in compaction and soil rearrangement to the side of the pipe.

Soil compaction or smear beneath the trenchless tine may occur at all working
depths, its extent being dependent on contact loads and pressures, soil
conditions and tine base width. Providing the compacted width does not
exceed approximately two pipe diameters, its effect on drain performance is
likely to be minimal. The major potential drainage problems are likely to
arise therefore, when trenchless tines are working below critical depth.

FACTORS INFLUENCING CRITICAL DEPTH POSITION

Two groups of factors influence critical depth position. One group is
associated with the implement, and the other with the soil.

Implement Factors

The width and inclination to the horizontal (rake angle) of the leading
share on the tine have a significant influence on critical depth position.
The narrower the share and the greater its rake angle, the closer the
critical depth will be to the soil surface. Commercially available trenchless tines fall into three classes as shown in Fig. 3. The rake angles of
the lower leading sections tend to decrease from type (a) to type (c) and
quite large geometry differences exist between the upper sections.
Comparative field tests with representative machines of similar share width
on two different soils, described in Spoor and Fry (1982b), showed them to
generate very similar soil disturbance patterns. The only difference at
depth was for the highly curved tines, type (c), to have slightly deeper
critical depths, (approximately 1 - 1.5 tine widths) than the other two
types. The widely different upper sections had no effect at pipe depth, but
they did influence surface disturbance.

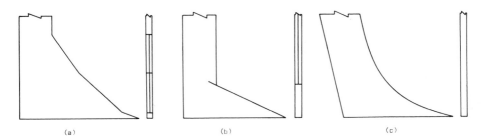

Fig. 3 Types of Trenchless Tine

The power unit tracks or wheels can influence critical depth position if the
soil disturbed by the tine prefers to heave under them. In these situations,
heave always breaks out at the inside of the tracks and critical depth is
shallower than if no track surcharge existed. This interference is greatest
at the greater working depths and with shares close to the track. Due to
their length, the highly curved tines, type (c), are frequently closer to the
tracks than the other types at any depth. For practical purposes, it can be
concluded that for tines of similar share width, there is little difference
in the nature of soil disturbance at drain depth between any of the tine
shapes, although critical depths may differ slightly.

Soil Factors

Whenever the resistance to upward soil movement is relatively low at working
depth compared to the lateral resistance, upward soil heave will dominate

resulting in above-critical-depth disturbance with few compaction problems.
In homogeneous soils, the resistance to upward movement increases with depth,
hence the change in disturbance at the critical depth.   Soils with stronger
surface layers or more compressible lower layers will tend therefore to have
relatively shallow critical depths and vice versa. High surface strengths
can result from surface drying and compaction.

Table 1 shows a representative range of the critical depth to tine width
ratios found during observations on approximately 100 field profiles generated by different machines under different soil conditions.   The figures are
presented solely as a guide to the order of magnitude of this ratio and it
must be stressed that considerable variation can occur even within the same
field as a result of changes in soil condition. To indicate the degree of
wetness or dryness at drain depth, representative values of dry bulk density
and ped density are quoted for the shrinking clayey soils, and % saturation
for the non-shrinking soils. Higher bulk density soils due to a lower
porosity are much drier than low density ones even when completely saturated.
The difference between bulk density and ped density is an indication of the
amount of drying and shrinkage within the profile.   The effect of soil
density on critical depth is very apparent across all soils.

SOIL CONDITIONS VULNERABLE TO DAMAGE WITH TINES
WORKING BELOW CRITICAL DEPTH

Drainage problems may arise with tines working below critical depth as a
result of lateral soil deformation reducing the size of the larger conducting
pores in the vicinity of the pipe.   These changes, with a corresponding
decrease in hydraulic conductivity, can occur as a result of soil compaction,
smear or simply soil shear without density change. The following characteristics and conditions have been identified as rendering a soil most
susceptible to reduction in hydraulic conductivity through the above
deformation processes:-

a. extensive lateral compaction - unsaturated low density single grain or
    fine textured soils and soils containing large air filled vertical
    fissures (unsaturated highly compressible soils).
b. smear or local soil reworking - higher density soils at high degrees of
    saturation.
c. extensive shear without density change - high moisture content, low
    shear strength, weakly structured fine textured soils, containing
    discrete macro pores in the form of old root or organism channels.

Soil conditions vulnerable to extensive lateral compaction were identified
in controlled soil bin tests using tines with depth/width ratios of 20:1,
see Spoor and Fry (1982a). In the sandy loam test soil, total volume
reductions of approximately 5% occurred to distances of about 4 tine widths
to the side of the tine at densities below $1.4t/m^3$.   Such volume changes
reduced hydraulic conductivity by approximately 50%, see Fig. 4. Under
fissured soil conditions, the extent of fissure or crack closure depended
upon crack width, the ease with which the soil around the crack could deform
and tine width. The cracks tended to fill initially by mass soil flow into
the cavity.

The extent of soil reworking or smear in the higher density soils at the tine
interface is limited to depths of approximately 2-10mm, the depth increasing
as soil shear strength decreases.   The hydraulic conductivity of a dense
clay fell from 4.5 to 0.02mm/day as a result of such smear.   Removal of the
5mm thick smeared layer returned the sample to its undisturbed hydraulic conductivity value.

Table 1. Representative critical depth/tine width ratios under different soil conditions.

Soil Texture	Soil Structure	Dry Bulk Density t/m³	Ped Dry Density t/m³	% Saturation	Overall Moisture Assessment	Ratio Critical depth / Tine width	Tine Width mm
Clay	Moderate to strong angular blocky or prismatic	1.5	1.8		Very dry	11-12	110-150
		1.5	1.7		Dry	8-9	110-130
		1.4	1.5		Moist	7-8	110-150
		1.3	1.4		Wet	6-7	110-130
Clay	Weak angular blocky	1.3	1.3		Very wet	3-4	110
Sandy Clay	Moderate angular blocky	1.5	1.7		Dry	8-9	110
	Weak angular blocky	1.2	1.2		Wet	6-7	130
Clay Loam	Weak angular blocky	1.5	1.6		Wet	6-7	130
Silty clay loam	Weak structure with discrete macro-pores	1.1		91	Wet	6-7	125
Sandy Loam	Single grain	1.5		65	Moist	10-11	110
Very fine sandy loam	Single grain	1.4		84	Wet	8-9	130
		1.55		93	Moist	11-12	190

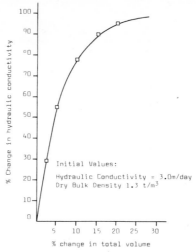

Fig. 4   Effect of Changes in Total Soil Volume on Hydraulic Conductivity
of a Sandy Loam Soil

Lower density fine textured soils at high degrees of saturation are extremely
difficult to compact but relatively easy to shear. In the absence of macro
pores, this shearing action has little influence on hydraulic conductivity.
Where, however, discrete macro pores are present, soil shear can readily
change macro pore shape, disrupt pore continuity and cause complete pore
collapse in certain cases. The results of this shearing action are
illustrated in Fig. 5, which shows the final distribution of macro pores
around a pipe, after trenchless installation just below critical depth, on a
92% saturated silty clay loam soil of $1.1 t/m^3$ density. The tine influence
can be seen to extend to approximately 150mm (1 tine width) either side of

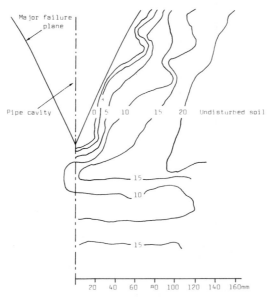

Fig. 5   Distribution of Macro-pores/unit of Plan Area, around
Bottom of Pipe Cavity. (Sample area = $0.004 m^2$).

the tine, the working depth was 0.9m. Figure 6 shows the significant
reductions in hydraulic conductivity resulting from the loss of different
numbers of macro pores in this soil.

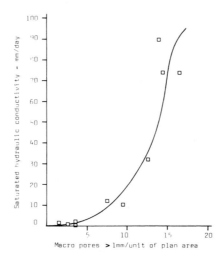

Fig. 6   Variation in Hydraulic Conductivity with Number of Macro
Pores/Unit of Plan Area.   (Sample area = 0.004m^2).

All lateral soil deformation problems will be made worse in situations where
tile box (or 'boot') width exceeds tine width.

PRACTICAL IMPLICATIONS OF DISTURBANCE PATTERNS
FOR DRAIN PERFORMANCE

The effect of trenchless installation on drain performance is dependent upon
the nature of the drainage problem, the type of disturbance and the prevailing soil conditions.   Few problems arise, whatever the type of disturbance,
in situations where a permeable backfill, either natural or artificial, is
used above the pipe as a connector for draining higher perched water tables
or removing water from mole drains.   Where this permeable zone is absent,
however, severe problems may arise.   This situation can occur with below-
critical-depth disturbance in certain low density, weakly structured fine
textured soils.   Perched water table drainage also benefits from the
increased loosening associated with above critical depth disturbance.
Excessive loosening above the pipe causes immediate mole channel collapse
where the mole crosses the drain line.   If, therefore, above-critical-depth
failure occurs, the disturbed soil must be recompacted before moling.   In
groundwater control situations, few side impedance problems are likely to
arise under any soil conditions, when the trenchless machines work above
critical depth.   Problems of impeded flow may arise under certain conditions
when the working depth is below the critical depth.

The soil bin tests showed unsaturated highly compressible soils to be
particularly vulnerable to side compaction below critical depth.   This type
of soil condition is much more prevalent in the surface layers than at depth,
in soils requiring drainage. At depth densities are usually above vulnerable
levels due to consolidation, and crack development is less due to lower
moisture deficits.   In most field situations, the vulnerable conditions will
be above critical depth and therefore of little consequence.   Serious
compaction problems are only likely to arise when these soil conditions are
overlain by stronger surface layers.

Smear problems at drain depth are very common after pipe installation on the finer textured soils under wetter conditions. They can occur due to tile box action, even when the leading tine is above critical depth. With the exception of the low shear strength clayey soils, smear is not usually complete along the pipe run and smeared layers usually crack on drying. Both of these features help to reduce the long term impedance effect of smear.

Loss of discrete macro pores, in any soil condition where they are the major water carrying channels, is a particularly serious problem, since new macro pores are usually slow to form. The extent of the pore losses will be greatest at high moisture contents, when shear strengths are minimal.

POSSIBLE MODIFICATIONS TO INSTALLATION TECHNIQUES TO MINIMIZE PROBLEMS

The risks of problems arising below critical depth are small on many soils. Nevertheless, if it is possible to work above critical depth, the risks can be reduced even further. Thus any modifications which will move the critical depth to below drain depth will be advantageous. This can be achieved within certain depth limits by increasing share width or by loosening the surface layers first, to reduce the resistance to upward soil movement. Care must be taken when increasing tine width to ensure that the critical depth is lowered below drainage depth. If this is not achieved, the wider tine working below critical depth will increase the extent of lateral deformation. This may cause further deleterious effects under vulnerable soil conditions.

A pre-ripping operation to approximately $\frac{2}{3}$ the required drain depth along the line of the drain before pipe installation, will help increase the critical depth by approximately 1 - 1.5 tine widths. Much greater benefits can be achieved through the use of suitably positioned, shallower working, forward leading tines, see Spoor and Fry (1982b), which will reduce if not always eliminate side compaction. Under the very vulnerable vertically fissured soil bin conditions, side compaction was reduced very significantly using these tines, although some partial crack closure still occurred. This technique has not as yet been tested in the low density, high moisture content, fine textured soils containing macro pores to assess its potential for reducing problems there.

In situations where side impedance cannot be avoided, Fry and Spoor (1982), in analogue studies, have shown that up to 90 - 95% of the impedance effects can be alleviated by part removal of the affected area, see Fig. 7. The critical area for side impedance removal is immediately alongside and if possible slightly below the pipe. Providing the impedance layers are of limited thickness, e.g. smeared zones and shallow compacted or macro pore loss areas, cutting blades attached to the rear of the tile box can be used to remove the impedance zone alongside the pipe. Cutting along a horizontal plane to produce a downward soil failure alleviates the problem without creating a further vertical smeared layer at the soil/blade interface. This method is less practical with thick impeded zones to the side of the tine.

The effects of compaction beneath the pipe on drain performance can be minimized by restricting the machine base contact width to approximately 2 pipe diameters. In circumstances where such a width restriction would cause below-critical-depth disturbance on a soil vulnerable to side compaction, the prevention of side compaction must always be given first priority. The placement of good quality permeable soil or artificial backfill in the leg slot above the pipe will improve the vertical connection for perched water table drainage. This may be necessary in certain low density, fine textured and unripe soils where impermeable zones can develop with below-critical-depth disturbance.

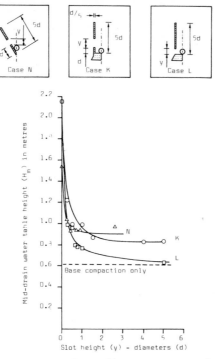

Fig. 7 Effect of Localised Smear Removal on Water Table Height at Mid-
drain Spacing (equivalent Drain Spacing 10m)

CONCLUSIONS

A major factor in the differing effects of trenchless and trenching
techniques on subsequent drain performance is the greater risk with the
trenchless technique of impedance layers developing in vertical planes
alongside the drain pipe.

Entry resistance for water flow into a drain pipe is very sensitive to side
impedance in groundwater control situations and mid-drain water table
positions will rise with increasing entry resistance.

Side impedance in surface, perched water table or mole drainage situations
will not affect drain performance since the major component of pipe inflow
is vertically downwards through the backfill material.

Significant side impedance is unlikely to occur with trenchless tines working above critical depth, but may develop in certain situations when working
below.

Three soil conditions have been identified as being most vulnerable to side
impedance with tines working below critical depth. These are:

a. unsaturated readily compressible soils.
b. those susceptible to smear or soil reworking at the tine/soil interface,
    and,
c. high moisture content weakly structured fine textured soils containing
    discrete macro pores.

The first set of conditions are not particularly widespread in the field at
depths below critical depth. Potential smear situations are very common.

Soils with discrete macro pores are confined to limited situations such as in certain fine textured marine sediments.

Techniques for surface loosening ahead of pipe installation are available to allow above-critical-depth disturbance over a wider range of soil conditions and working depths. Side impedance zones of limited extent can readily be removed during installation by blades fitted to the rear of the tile box. Vertical permeability problems in the leg slot area can be reduced by adding artificial backfill or soil from the most permeable horizon.

With care, and minor modifications in certain cases, trenchless installation techniques can be used with confidence over a wide range of soil conditions and drainage situations. Under these circumstances the risks of entry resistance problems, and hence high mid-drain water tables developing as a result of impedance at the side of the pipe are small.

The most serious problems are likely to arise with extremely deep drainage and on high moisture content, weakly structured fine textured soils containing discrete macro pores. Further work is required in these situations to investigate possible modifications to the technique to minimize or overcome the side impedance problem.

ACKNOWLEDGEMENTS

The authors wish to acknowledge the financial support provided by the Agricultural Research Council, London, England, and the willing support of many drainage contractors in Canada and the United Kingdom.

REFERENCES

1. Cavelaars, J.C. 1967. Problems of water entry into plastic and other drain tubes. Symposium, Institution of Agricultural Engineers, Silsoe, England. Paper No. S/E467. 13pp

2. Eggelsman, R. 1979. Comparisons between trenchless and trenching sub-drainage. Proc. Int. Drainage Workshop, Wageningen, The Netherlands. Ed. J. Wesseling. Publ. ILRI, 536-542.

3. Fry, R.K. and G. Spoor. 1982. The influence of soil compaction and smear in the vicinity of a pipe drain on drain performance. J. Agric. Engng. Res. (In Press).

4. Godwin, R.J. and G. Spoor. 1977. Soil failure with narrow tines. J. Agric. Engng. Res. 22, 213-228.

5. Naarding, W.H. 1979. Trench versus trenchless drainage techniques. Proc. Int. Drainage Workshop, Wageningen, The Netherlands. Ed. J. Wesseling. Publ. ILRI, 543-560.

6. Olesen, S.E. 1979. Trenchless drainage experiments in Denmark. Proc. Int. Drainage Workshop, Wageningen, The Netherlands. Ed. J. Wesseling. Publ. ILRI, 527-535.

7. Spoor, G. and R.K. Fry. 1982a. Soil disturbance generated by trenchless drainage machines and low rake angle narrow tines. J. Agric. Engng. Res. (In Press).

8. Spoor, G. and R. K. Fry. 1982b. Field performance of trenchless drainage tines and implications for subsequent drainage system performance. J. Agric. Engng. Res. (In Press).

TWO DECADES OF EXPERIENCE
WITH DRAINAGE FILTERS IN THE
FEDERAL REPUBLIC OF GERMANY

By

RUDOLF EGGELSMANN*

In 1960 the first corrugated PVC-drain-pipes were installed in
Northwestern Germany.
In the Soiltechnological Institute' Bremen we developed a
publication entitled "Documentation of Modern Subdrainage
Materials" for information, education and research (EGGELSMANN,
1982) to respond to questions concerning the use of corrugated
PVC-drain-pipes.

PRODUCTION OF CORRUGATED PVC-DRAIN-PIPES

From 1965 to 1970 in the Federal Republic of Germany (FRG) the
production of corrugated PVC-drain-pipes nearly doubled to
80 million meters per year. (Table 1).

Table 1. Annual German Production of Corrugated PVC-Drain-Pipes

Year	Million Meters of Pipe
1970	44
1971	62
1972	60
1973	70
1974	70
1975	75
1976 - 1981	75-80

The entire production of corrugated PVC-drain-pipes since 1960
has amounted to about 850 million meters. Nearly 65% of the
production were pipes with a diameter of 50 mm, which in FRG
is the "normal" drain pipe diameter by DIN 1185 (German
Standard).

The corrugated wall of the pipe has improved in stability and
the quantily of PVC material used has been reduced since
corrugated PVC-drain-pipes was introduced. Complaints about the

*Dr. rer. nat. R. Eggelsmann, M.E., C.E., Professor
 Geological Survey of Lower Saxony -
 Soiltechnological Institute Bremen
 Friedrich-Missler-Str. 46-50, D-2800 Bremen 1 FRG

longevity of the corrugated PVC-drain-pipes have been less than 1% in more than 20 years.

DRAIN FILTERS

The use of machinery for drainage installations in Western Europe has increased the use of drainage filters. This is because of the greater use of drainage machinery in moist soils and unfavorable weather conditions as well as the action of the trenching chain.

We have to distinguish between pipe filters and trench filters. A pipe filter covers a drain pipe either partly or completely and is laid either at the same time as the drain pipe or is added afterwards. It also protects the bottom of the pipe. A trench filter is put into the trench to a certain depth. It is used particularly for interceptor drains and in connecting mole drains with pipe drains.

At the present stage of knowledge drainage filters should fulfill the following functions either separately or together:

- mechanical effect, i.e. separation of soil particles over 0.02 mm. The finest particles, clay and fine silt, should pass through the filter.
- hydraulic effect, i.e. reduction of the entrance resistance at the drain pipe. The permeability of the drainage filter should be at least ten times greater than that of the soil near the pipe.
- biochemical effect, i.e. decrease or avoidance of ochre clogging of the drain.

The question of where and when is a drainage filter necessary only can be answered according to the characteristics of the soil and the methods if installation. Whether primary silting which occurs immediately after installation and lasts a short time, or secondary silting which lasts continuously must be considered. Drainage filters are necessary particularly if the soil has a strong tendency to silt. Also it may be necessary when trenching chains are used and when drains are installed in moist soil (Table 2).

Table 2. Class od Soil Versus Tendency to Silt (KUNTZE, 1974)

Soil Conditions	Primary Silting	Secondary Silting
coarse sand	-	-
medium sand	+	+
fine sand	++	++
silty sand	+++	+++
loamy sand	+	-
clayrous sand	+	-
silt	+++	+++
sandy silt	+++	+++
loamy silt	++	++
sandy loam	+	-
silty loam	++	-
loam	+	-
sandy clay	+	-
loam clay	+	-
clay	-	-
low decayed peat	-	-
medium decayed peat	-	-

```
- no
+ low
++ medium      tendency to silt = need of filter
+++ strong
```

The mechanical effect if a filter is largely dependent on the
porosity of the filter, which in turn is influenced by the
density of the packing. Furthermore it depends on the type,
thickness and the composition of the fibers. Drainage filters
with large pore spaces have shown little puddling because of
the discharge velocity at the drainage filter - soil interface.

FILTER MATERIALS

Organic filter materials which have been used are gravel,
peat, straw, brushwood, sawdust and coco-fibre. Synthetic
filter materials include fibers, foam plastics (styro foam,
Polyurethan), fleeces, felt, and slag. Table 3 contains
the most important data of 12 different drainage filters
which were examined by Burghardt (1976).

Table 3. Physical Properties of Drain Filter

Material	Thickness of filter cm	Density g/l	Density g/cm^3	Pore Volume %
Rye straw	1.4	64	1.58	96
Rye straw	1.3	96	1.58	94
Coco-fibre	0.9	107	1.52	93
Wood-fibre	1.2	90	1.50	95
Fibrous peat	0.85	66	1.56	96
Acrylic-cellulose mixture	8.33	52	1.45	96
Polyester fibre	0.5	138	1.38	90
Polypropylen	0.1	202	0.90	78
Polypropylen	0.35	118	0.91	87
Glass-fibre	1.3	57	2.62	98
Glass-fibre fleece	0.03	168	1.93	81
Polyurethan foam	1.2	39	1.14	97

The diameters of the pores of a filter limit the mobility of the various particle sizes of the soil. In the case of natural fiber materials like oat and rye straw, wood-fiber and coco-fiber, decay was observed in laboratory and field tests. The amount and speed of the decaying process was shown to be largely dependent on aeration and moisture content. The nutrient content of the drainage waters also influenced the decaying process. In the case of straw filters which were constantly under water, development of bacteriological slime was observed. These slimes lowered the permeability of the filters to unacceptable levels.

OCHRE CLOGGING

Drainage water may contain reduced and therefore soluble iron or manganese compounds. Within and at the drain pipe the iron dissolved in the drainage water gets in contact with oxygen. Because of oxidation and precipitation, a gelatinous iron containing silt develops and becomes powdery after drying. This process called ochre clogging may block the drain pipe rapidly and affect the efficiency.

In the course of extensive laboratory and field experiements it was discovered that organic filter materials containing tannin formed relatively stable compounds which slowed down the chemical and biochemical development of ochre. Tannin containing substances are found chiefly in the shavings of various species of wood such as in the bark of mimosa (Acacia).

In a experiment which lasted more than five years, sawdust con-

sisting of 90% conifer and 10% decidious wood was tested along
with straw, glass wool and other materials. The drainage filter
made of sawdust significantly slowed down ochre clogging, thereby lowering the entrance resistance of the water at the drain
pipe. Consequently, the hydraulic efficiency of the entire
drainage system was improved. In contrast, the other drainage
filters (for instance glass wool) and the entry holes located
in the drain pipes become blocked by ochre.

Intially after installation, the drainage water of a drain with
a saw dust filter (or Antoc-filter) contains larger amounts of
organic substances and phenols. They affect the biology of the
outlet channel. Therefore a dilution of 1:10 to 1:20 is
desirable.

During the past few years flexible ready-made filter pipes have
been installed in the FRG. In Northwest Germany these pipes
accounted for more than 60% of the installations since 1976. Rye
straw and coco-fibre are used primarily. The use of fullfilter
pipes has increased in the Netherlands, Belgium, France and
Denmark.

COSTS

The cost of whole coils of the most important ready-made PVC-
filter-pipes in Northwest Germany plus value added tax is shown
in Table 4. Trench filters of gravel or granulated slag cost
about 4 to 6 DM/m or more.

Table 4. Cost of PVC Ready-Made Drain Filter Pipes in Northwest
Germany (1981)

$\emptyset$ (mm)	DM per 100 m	
	coco	straw
50	80	70
65	120	110
80	210	180
100	300	280
125	470	400
160	750	620
200	1400	(900)

For several years in Northwest Germany almost all drainage outlets have been made of rigid PVC. They are one meter long and
have a latticed backwater gate. The costs range from 13 DM a
piece ($\emptyset$ 50) to 75 DM a piece ($\emptyset$ 200) plus value added tax.

Settling basins, 0.8 m in diameter and 1.50 m in length, and
inspection wells, 1 m diameter and 2 m in length, are made of

concrete. They each cost about 150 or 250 DM.

CONCLUDING REMARKS

Drainage materials have had a positive effect in many ways in the FRG. They have revolutionized drainage technology through the use of less personnel. Thus, the costs have been lowered and the efficiency has been increased. As a result the costs for pipe drainage have been stable for the last few decades. Sometimes the prices even declined. New drainage materials have improved the efficiency of pipe drainage. This is especially true for drainage filters.

REFERENCES

1. R. EGGELSMANN: Subsurface Drainage Instructions. -
    Ist German Ed. (with English and French subtitles and summary), Verlag Wasser und Boden Hamburg 1973.
    Ist English Edition (translated by U.S. Department of Agriculture), Verlag Paul Parey Hamburg 1978.
    Ist Russian Edition Kolos Moscow 1978.
    2nd German Ed. Verlag Paul Parey Hamburg 1981.
    Other foreign Edition in preparing.

2. R. EGGELSMANN: Documentation of modern subdrainage materials. -
    Z.f.Kulturtechnik u. Flurber. 1982 (German in printing).

3. DIN 1185 (German Standards) - Subdrainage as control of drainage, and subsoil improvement. - Ed. Beuth Verlag Berlin 1973 (in German).

4. DIN 1187 (German Standards) - Plastic drain pipes made of PVC hard. Ed. Beuth-Verlag Berlin 1980 (in German).

CURRENT CONSTRUCTION TECHNIQUES
EFFECTS ON DRAINAGE SYSTEMS IN ONTARIO

R. W. Irwin   J.R. Johnston
Member, ASAE   Member, ASAE

The land drainage industry has survived and prospered through a very rapid
change in technology over the past 15 years. We propose to examine this
industrial revolution and establish how changes in the drainage construction
industry has had an effect on Ontario on-farm drainage systems. Changes
have occurred in pipe material, installation machinery and techniques and in
system layout and design. Each of these areas will be compared for drainage
system efficiency.

Drainage Pipe

Fifteen years ago the principal drainage pipe in use was clay drain tile.
In Ontario this material was produced from shallow surface clay pits. There
were 31 plants producing about 20 million metres of pipe annually. Irwin
and Spencer (1970) examined the industry and found the product quality to be
variable. The tile barely met the minimum ASTM standard for strength and
did not meet the freeze-thaw test. Installations used glazed sewer pipe in
frost prone situations such as at catch-basins. Annual growth in the clay
tile industry from 1966 onward was about 10 percent. Generally this is
attributed to shale plants coming into the market. Minimum return on
investment was the chief cause of a rather stagnant clay tile industry.

The demand for clay tile exceeded the supply in 1966-67 and other products
such as concrete tile and bituminous fibre pipe were examined. A modest
increase in production occurred but did not fill the demand. In 1967
Advanced Drainage Systems introduced corrugated plastic drainage tubing to
North America on a commercial basis. The following year a group of Ontario
contractors formed the Big O Drain Tile Co. to manufacture corrugated
plastic drainage tubing in Canada. Farmer acceptance was not immediate,
extension agents and researchers were wary of the new product. However,
contractors found the new pipe had many advantages and its use expanded
rapidly.

Figure 1 shows the increase in use of drainage pipe in Ontario from 1965 to
1980. Over the period the total use has nearly tripled as farmers strive to
maximize their return on investment. The annual rate of increase is about
11 percent. The change in use of pipe product material is also shown in
Figure 1. Plastic tubing has shown a steady growth which has filled the
demand of the industry and has taken the bulk of the market from the clay
tile industry. Concrete pipe has a minor part of the drainage market.

Fifteen years ago the bulk of the clay drain tile was laid by hand using a
tile hook. The contractor did the quality control during the tile laying
operation. Because the soil-conditions were visible all the time, suitable
crack widths were selected by the drainage contractor during the laying
operation. This important operation generally guaranteed the farmer a good
job. There were a few problems in failure of the clay tile due to excess
surface loading. These failures were detected readily from the surface and
were repaired easily.

The authors are: R.W. Irwin, Professor, School of Engineering, University of
Guelph, Guelph, Ontario and J.R. Johnston, Drainage Co-ordinator, Capital
Improvements Branch, Ministry of Agriculture and Food, Toronto, Ontario.

Today, drainage systems generally are constructed with automatically laid corrugated plastic drainage tubing. The strength of the plastic tubing is realized from the surrounding soil using the principles of flexible pipe theory vs the rigid pipe theory that must be used for clay drain tile design. The inlet water area exceeds that of clay drain tile and should result in improved soil-pipe hydraulic entrance characteristics. The internal hydraulic resistance is greater and results in a discharge capacity which is 70-75% of that clay drain tile but this characteristic is of little importance except for main drains. Stretch during installation adversely affects the strength of plastic tubing. Today most Ontario installation equipment has power feeders to minimize stretch.

Drainage tubing manufacturers and drainage contractors tend to use the same internal pipe roughness for all types of pipe. This practice creates some surcharging of the main drain during peak flows.

In time, sunlight can deteriorate the plastic material rendering it unsuitable for use. Plasticizers are incorporated in the resin mix to inhibit this deterioration.

In spite of what some people saw as questionable strength and durability, plastic drainage tubing has been accepted by the farmers as a satisfactory drainage material. It now boasts a 15-year history of satisfactory performance. The authors conclude that plastic tile, properly installed, will last probably as long as clay tile and will perform the same function in the field equally well.

Machinery

Fifteen years ago the principal drainage machine used by contractors was the wheel trencher. Irwin (1972) compared the various machines in use and stated the forward digging speed depended on the tile laying operation but seldom exceeded 550 m/h. With a five man crew, contractors could average about 1500m installed per day. The deeper the installation the slower the rate, however, price seldom reflected this fact.

The excavated material was windrowed beside the trench for backfilling with a road grader. A few machines had conveyor belts to transport the excavated material to the back of the shoe for automatic backfilling. Today the backfilling practices are not significantly different but increased power allows digging speeds approaching 730 m/h with corresponding efficiencies in installation cost.

About 1965 chain trenching machines were introduced from Europe. Using a rapidly moving, toothed chain these machines open a trench about 250mm wide. Excavated material is augered to both sides of the trench for backfilling with a farm tractor and blade. The daily production could be increased to about double that of a wheel trencher, i.e. 1100 m/h. Maintainance costs were high and these trenchers proved unsatisfactory in the many stoney clay till soils of Ontario. Today a few chain trenchers are still in use in those areas that are predominately stone free clay soils. To quote one contractor "anything that rattles wears and costs money."

The first drainage plow was introduced in Ontario in 1968. The machine opens a slot by shearing and lifting the soil above the desired tile depth. Tubing is automatically fed down a hollow shoe and the soil closes over it leaving a ridge about 0.6m in height. Extension agents and researchers were concerned about the quality of installation resulting from this new concept. The original machines were very expensive but by 1974 a new generation of less expensive machines was being marketed and competition was established. The drainage plow increased the rate of installation to about four times that of the wheel trencher. Contractors can average 45-7500 m/d.

Today there are 6 drainage plows laying clay drain tile in Ontario while the majority (over 100) lay plastic drainage tubing. It is the availability and acceptability of this tubing which has made the drainage plow feasible. The chief advantage of the drainage plow to the contractor is the substantial reduction in maintenance cost and increased productivity. The farmer receives a rapidly installed system with few stones brought to the surface. The performance characteristics for drainage plows were discussed by Irwin and Johnston (1978).

Figure 2 shows the number of drainage machines by type licensed in Ontario. The number of wheel trenchers is still high, however on a production basis, drainage plows now install over 60 percent of all the agricultural drainage pipe.

The type of drainage machine affects the farmer in a number of ways. First, the average drainage job of 12 ha, or 7500m would take a wheel trencher over a week to complete, a drainage plow would complete the job and be off the farm in less than two days. The farmers time is not committed to the drainage operation for as long.

The drainage plow has kept the cost of a drainage installation to a minimum due to its greater work capability and lesser labor cost.

Not all plows are equal. Some must install drains shallower in order to work faster. Contractors should recognize that speed comes from organization and experience, not with the plow itself.

Grade Control

Grade control on a drainage machine is essential for quality work. In 1965 grade control was by grade stake and sighting across cross-arms by the eye of the machine operator. The results were acceptable but this was primarily due to the slow forward speed of the wheel trencher.

With the introduction of the drainage plow in 1968 the increased forward speed demanded an improved system - grade stakes were too slow. A surveying level with radio control was used for a couple of years to control grades but was not efficient in labor nor really accurate in results.

Lasers were applied to grade control in 1965 and have been adapted to all types of drainage equipment. The system is expensive but saves in labor cost and time of specialists in the field. It is as accurate as the expertise of the operator and general maintenance of the machine and equipment will permit.

Laser is used by contractors to survey the farm for drainage. This is one of the less satisfactory results of the application of technology. Contractors tend to eyeball the land and estimate the grade. This is then dialed into the laser command post and checked for depth. If the depth at the critical point is too shallow for the drainage pipe the grade is changed for that point. In land where there is some undulation this survey technique results in lateral drains which are often too deep in places because most laterals are designed for average grade conditions. The laser system will permit changes in grade but the implementation is not simple.

Properly applied each lateral is surveyed, grades calculated and the survey results given to the operator who will dial in the selected grades for his machine to perform. Installations from this method allow uniform depth drainage in undulating lands just as they would have been designed by a contractor using grade stakes and a sight rod.

Figure 3 shows the adoption of laser grade control in the industry. Laser is a highly accurate and good method of grade control. The results are more accurate grade control today than 15 years ago.

Drain System Layout

Drain system layout and field installation is the major place where technology has changed the on-farm situation. Many of these changes are improvements. Today most installations are systematic with random lines used primarily in undulating heavy clay soils with low permeability.

Outlet pipe with rodent guards are now used on every installation. This was not the case 15 years ago. Now outlets are available generally with sufficient depth although contractors tend to install outlets at the bottom of ditches with little concern for the results.

The average drainage density has increased over the years. The closer spacing is a combination of need for improved drainage due to higher valued crops, heavier and larger field equipment, soil degradation and promotion by drainage contractors and tile salesmen.

Lateral drains installed with a plow must begin from a "start hole" at the main. Digging the start hole often delays the work when lateral drains are short. Contractors often orient the drainage system to have as few connections as possible. In many instances the orientation of the lateral drain is not suitable to the present system of farming and crop layout. For example, farmers prefer to plant row crops on the long way of the field, therefore, the drainage should be on the short way of the field. This would make more connections for the contractor and therefore usually is not done.

There are many more tile-soil problems today. Drainage envelope protection is used on nearly 15 percent of the work. Fifteen years ago its use was on about 4 percent of the work. The new sock type filter material applied to drainage pipe at the factory is far superior to the rolled fibreglass installed in the field 15 years ago. While the material is better there is considerable doubt regarding need for such a high scale of use. Unnecessary use of synthetic filters increases the cost of the work by 15%. It appears that contractors recommend filter for their financial protection rather than for drain protection. To a great extent this is because insufficient investigation of the soil is made in advance. When a drainage plow is used the operator is unable to see the soil conditions at drain depth and therefore is uncertain of the variation in soil conditions at drain depth. It may be that excess protective sock is used to overcome these uncertainties.

In dry fine textured soil drains installed with a plow are reported to respond faster the first year due to the subsoiling effect. The depth of settlement of the disturbed surface also is reduced.

Old drainage systems tend to be shallower than systems installed now. When drainage plows cut these old systems this is usually noted and the tile are repaired or connected to the new system at time of installation. However one contractor states that he spends two weeks each spring repairing old systems where wet spots remain in the field and connections were not made.

Because of the greater depth, wheel and chain trenchers hit more stones, thus reducing the rate of work. The drainage plow has a greater capability with stones and its speed is not affected as much although dryness, wetness and large stones often affect the depth to which lateral drains are installed.

High powered new equipment with lower ground pressures works earlier in the spring and later in the fall when the soils are wet. It is the authors' opinion that the farmer does not get as satisfactory a drainage system as compared to those installed at drier times of the year.

Farmers now clear new land and improve land that in former years would have been reserved for unimproved pasture. Drainage of these soils results in more problems with quicksand and iron ochre. The spreading of large quantities of liquid manure enriches the soil amd can cause bacterial growth in drains leaving them totally blocked. Instances of this type of blockage have increased over the past 15 years.

The blade on a drainage plow must be changed for laying larger diameter main drains. Changing the blade takes time and labor. Thus, where the length of main drain is short there is a tendency to modify the system design so the blade change is unnecessary. This often results in twin main drains side by side, for example two 150mm mains instead of an 200mm main. Often the farmer decides the size of main drain based on his ability to pay. So instead of the twin 150mm the contractor may in fact install only one 150mm main where previously an 200mm would have been used.

CONCLUSIONS

1. Corrugated plastic drainage tubing is used for lateral and main drains on 82% of the installations.

2. Contractors and tubing suppliers tend to use the same internal roughness for selection of main drain diameter independant of the type of pipe. This results in surcharging some drains when design capacity is reached.

3. The quality of the drain pipe has improved over the past 15 years.

4. Contractors are better trained today, and have more responsibility and consequently do a better job.

5. Installation time on a farm is so short that the farmer does not have an opportunity to consider other needs before the contractor has left.

6. The farmer is no longer associated intimately with the job and may miss problem areas which in other times would have been seen by him (e.g. large stones, quicksand, old drains, etc.).

7. Grade control of the drains is more accurate today.

8. Drainage plows bring fewer stones to the surface for the farmer to pick.

9. When new drains are installed using a drainage plow. Old drainage systems tend to create problems because often they are not connected to the new system.

10. Drains are installed too early and too late in the year. This practice damages many soils because of the greater soil wetness during these periods.

11. There are more systematic drainage systems and fewer random systems being used.

12. There is a tendency toward closer drain spacings and increased depth.

13. Outlet pipes with rodent guards are now in general use.

14. More polyester sock is used today.

15. There is a tendency to reduce the diameter of main drains or to alter the design so there may be multiple outlets. This creates more maintenance for the farmer.

16. Often drains are oriented, in a direction advantageous to the contractor rather than the farmer.

17. Drainage plows have kept the cost of drainage to a low level.

REFERENCES

1. Irwin, R.W., V.I.D. Spencer. 1970. The Clay Drain Tile Industry in Southern Ontario. Can. Agr. Eng. 12: 107-109.

2. Irwin, R.W. 1972. Drainage Equipment - What Type for the Future? ASAE Paper NAR72-201.

3. Irwin, R.W., J.R. Johnston. 1978. Performance of Drain Plows in Ontario. ASAE Paper 78-2044.

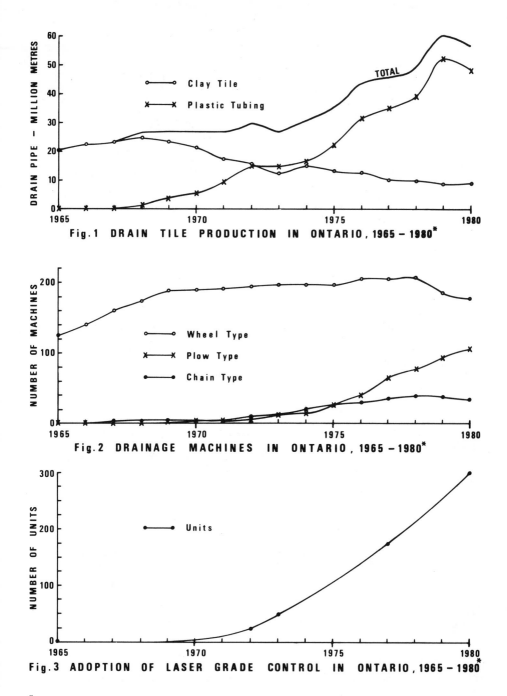

Fig.1 DRAIN TILE PRODUCTION IN ONTARIO, 1965 – 1980*

Fig.2 DRAINAGE MACHINES IN ONTARIO, 1965 – 1980*

Fig.3 ADOPTION OF LASER GRADE CONTROL IN ONTARIO, 1965 – 1980*

*DATA FROM CAPITAL IMPROVEMENTS BRANCH, OMAF

MOVEMENT AND RETENTION OF WATER IN A HEAVY CLAY SOIL[1]

author block## H. M. Selim[2], C. E. Carter[3], and E. D. Aviles[4]

Little is known of the movement and retention behavior of water in heavy clay soils. This is due primarily to their low permeabilities, poor drainage, and in some cases swelling and shrinkage characteristics. Heavy clay soils in the Southeast region frequently have a permanent and shallow water table throughout the growing season. In addition, during periods of frequent rainfall the water table often inundates the soil surface. Soils high in montmorillonitic clay content often exhibit extensive swelling and shrinkage on wetting and drying. Very little effort has been made to quantify the soil physical characteristics which influence the water flow pattern and water retention in these clay soils.

The purpose of this study was to characterize the soil water behavior in Sharkey clay soil in southern Louisiana. Laboratory measured water retention and saturated hydraulic conductivities are discussed. Field measured water movement under drainage and evaporation conditions is also presented. Attempts were made to determine the in situ hydraulic conductivity for this soil.

THEORETICAL ANALYSIS

Two methods were used to calculate the in situ soil unsaturated hydraulic conductivity. The first was the instantaneous profile method (commonly known as the drainage method) in which water redistribution in the soil profile is assumed to be due primarily to vertical downward water flow or drainage where evaporation and lateral water flow are ignored. This method relies on the basic form of Darcy's law in one dimension

$$v = Q/At = -K \, dH/dz \tag{1}$$

where v is the water flux (cm/hr), Q the volume of water (cm^3) passing through a cross sectional area A (cm^2) and time t (hr), K the hydraulic conductivity (cm/hr), H the hydraulic head (cm) and z the soil depth (cm). The method of calculating the hydraulic gradient term (dH/dz) is an approximate one where an average gradient is estimated over an incremental depth Δz and a time step Δt. If i denotes incremental distances and n is the time step (i.e. $\Delta z = z_{i+1} - z_i$ and $\Delta t = t^{n+1} - t^n$), then

$$dH/dz = [(H_{i+1}^{n+1} - H_i^{n+1}) - (H_{i+1}^n - H_i^n)]/(z_{i+1} - z_i) \tag{2}$$

where the hydraulic head H_i^n indicates the hydraulic head at the depth i and time of redistribution n. Also the hydraulic head was considered as

$$H = h - z \tag{3}$$

where h is the soil water pressure head (cm) and z is depth with downward considered positive.

---

publication info[1] Contribution from the Agronomy Department, Louisiana Agric. Exp. Station, Louisiana State Univ., and USDA-ARS, Baton Rouge, La. 70803.
[2] Associate Professor.
[3] Agricultural Engineer, USDA-ARS.
[4] Research Assistant.

128

In order to calculate the flux v, a unit cross-sectional area A was
assumed and the volume of water passing through some point z in the soil
is calculated from the changes in the volumetric soil water content
(cm^3/cm^3) over a time period Δt. Because evaporation is assumed to be
zero, the changes in water content distributions over a time period Δt
can be obtained by integration from the soil surface to the desired
depth z,

$$Q/A = \int_o^z (\theta^{n+1} - \theta^n) \, dz \tag{4}$$

The integration was carried out by a simple summation method with depth
and time. The soil water contents in the above calculations were
obtained indirectly from the soil-water retention (θ vs h) curves. For
further details on this instantaneous profile method, the reader may
refer to Watson (1966) and Cassel (1975).

The second method of calculating in situ soil hydraulic conductivity is
the evaporation method (commonly referred to as the plane of zero flux
method), (Arya et al, 1975). The main advantage of this method is that
no surface seal is required and water redistribution in the soil profile
occurs under evaporation and drainage simultaneously. The method relies
on defining a plane of zero flux within the soil profile at all times for
the purpose of calculating water flux (upward or downward). Under
evaporation conditions, the plane of zero flux is a moving boundary and
penetrates to lower soil depths as soil drying continues. Above the
plane of zero flux the movement of water is upward due to evaporation and
below it the movement is downward due to drainage. The potential
gradient dH/dz, which is obtained from the tensiometer readings with
depth, defines that plane and the summation of water content changes over
time is carried out from that depth to the point of interest in order to
obtain the water flux at a desired depth z.

MATERIALS AND METHODS

A field site was selected near St. Gabriel, Louisiana as a representative
Sharkey clay soil. The soil was fallow for the two years prior to the
initiation of this experiment. The soil topography was nearly level and
slightly convex. Three adjacent plots (2.2 by 2.2 m) were instrumented
on this soil. A trench 15 cm deep was excavated around the perimeter of
each plot. A wooden dike, constructed of four boards, each 2 cm thick
and 2.2m long, was installed in the trench around each plot. The corners
of the dikes were sealed to prevent water leakage. The trenches were
backfilled with the excavated material leaving the dike extending 15 cm
above the soil surface. Three sets of tensiometers (porous ceramic cups
connected to a mercury manometer) were installed in each plot at depths
of 10.2, 20.3, 30.5, 40.6, 50.8, 61, 71.1, 81.3, 91.4, 106.7, 121.9, and
152.4 cm from the soil surface. In addition, two tubes were installed
in the middle of each plot.

Water was applied to each plot from a 500 gallon tank for a few days. The
water remained ponded on the soil surface for some 5 days. When the
tensiometer readings indicated that the entire soil profile (up to 152
cm) was close to saturation, water application to the plots was
terminated. After water application was terminated, and the ponded water
had infiltrated, the soil surface was covered with polyethylene sheets to
prevent surface evaporation. In addition, plot covers (polyethelene
sheets attached to a wooden frame) were placed on each plot for the
prevention of infiltration due to rainfall as well as evaporation from
the soil surface. Tensiometer readings were monitored daily during
May to December, 1981 and May through June, 1982. As will be discussed

in the Results and Discussion section, this method of calculating the in
situ hydraulic conductivity proved unsuccessful.

Due to the swelling and shrinkage behavior of this soil, its low
permeability, and frequent rainfall, the water table remained shallow and
often inundated the soil surface. In addition, the presence of zero flux
boundary at the soil surface (i.e. no evaporation) resulted in an
extremely slow change of tensiometer readings with time and depth under
drainage conditions. As a result, a wide range of soil suctions and
water contents could not be achieved. Because of these difficulties, the
surface insulations were removed and water movement during simultaneous
water drainage and evaporation was monitored. This method, which is
referred to as the evaporation or zero plane of flux method, proved to be
successful in obtaining a drying cycle of the soil moisture profile
during periods of no rainfall.

Gravimetric moisture contents with depth and the depth of water table
were measured frequently in each instrumented plot. These data were
used to correlate piezometer measurements with pressure head
measurements in determining the location of the water table with time.
Gravimetrically obtained soil moisture content distributions were
correlated with that obtained indirectly from the soil moisture retention
or release curves.

In an area adjacent (6 m) to each plot, a pit was dug to a depth of 180
cm in order to obtain a profile description and samples from each soil
layer. Several replications of undisturbed soil core samples (5.4 cm
diameter and 3 cm long) were taken at approximately the middle of each
horizon. These undisturbed cores were used for the moisture content vs.
suction determination using Tempe-cells (Soil Moisture Equipment Corp.,
Santa Barbara, CA) as described by Wilson (1981). The soil water
pressures applied were 0, 20, 50, 100, 150, 200, 300, 500, 750, and 1000
cm of water. Another set of samples (6 cm rather than 3 cm long) were
also collected from each layer to determine soil bulk density. A third
set of undisturbed core samples (7.6 cm in diameter and 5.1 cm long) were
used to determine the saturated hydraulic conductivity in the laboratory.
These somewhat larger cores were used in order to obtain a representative
soil volume for hydraulic conductivity measurements. The procedure used
was that of the constant water head method for measuring saturated
conductivity (Klute, 1965). For further details on soil core preparation
and the saturation of the soil, the reader may refer to Wilson (1981).

RESULTS AND DISCUSSION

From the three soil pits, which were dug adjacent to the instrumented
plots, a profile description was made. The soil was comprised of 5
layers; Ap (0-17.8 cm), A1 (17.8-43.2 cm), AC (43.2-78.7 cm), C1
(78.7-106.7 cm) and C2 (106.7-160 cm). The structure, texture, and color
properties were similar to those described in the Soil Survey of East
Baton Rouge Parish (1968).
The soil (dry) bulk density of the Sharkey clay soil at various depths
are shown in Fig. 1. The results indicate a slight decrease of bulk
density with depth. More important is the greater variability of bulk
densities in Ap, A1 and AC horizons in comparison to lower soil layers.
The presence of extensive shrinkage cracks in this soil is the primary
reason for such soil bulk density variability in the top 60 cm of the
soil profile. The fact that low bulk density values (less than 1 g/cm^3)
were encountered in these layers is an indication of high pore space
resulting from the shrinkage and swelling behavior of this clay soil. It
is interesting however, to note that there appeared to be no significant
differences in density among the three plots.

Results of the laboratory measured saturated hydraulic conductivity
(K_{sat}) with soil depth for the three instrumented plots are shown in
Fig. 2. Variability of K_{sat} for all plots ranged between 2 to 3 orders
of magnitude. It should be noted here also that unrealistically high
values of K_{sat} for all soil layers were obtained. Such high values were
a direct result of water channeling through continuous shrinkage cracks
along the flow path in the soil cores (see Philip and Thomas, 1979). In
fact, several samples were discarded due to their extremely high flow
rates. This was particularly true for the surface horizon where out of
18 cores (6 from each plot) a total of 14 samples were discarded. It
should be mentioned here that Lund and Loftin (1960) reported zero values
for the water permeability in the Ap and C layers of a Sharkey soil
obtained from a different location.

Results of the soil water content-suction relationships for all three
plots are presented in Fig. 3. These results represent the average data
for each plot at the various suction values. For this clay soil, the
results indicate no definite air entry value. In addition, the decrease
of soil water content was gradual over the entire suction range applied
(0 to 1 bar). The soil water content values at zero suction, as
indicated on the ordinate in Fig. 3, correspond very closely to the
total porosities which were based on bulk density values. This is an
indication that full saturation of the soil core samples was achieved
prior to the application of the incremental pressures. The results shown
in Fig. 3 also indicate significant variability among the three
experimental plots. Such variability was not restricted to the top
layers but extended also to lower depths.

During the course of the laboratory determination of the soil water
retention curves shown in Fig. 3, it was observed that the samples in
the tempe cells shrank at suction values of 100 to 300 cm. This change
in volume probably resulted in the soil's loss of contact with the
porous plate. For this reason, the dimensions (length and diameter) of
the soil samples were measured at the termination of the experiment.
These results are given in Table 1 along with the change of volume due
to the shrinkage of the samples. This was sought in order to give an
indication of the soil shrinkage as the soil water content decreased due
to increased suction. The change of volume ranged from 22 to 39% of the
original volume. Such excessive shrinkage was more pronounced in the
lower soil depths (below 70 cm), however. Such data provides actual
measurements (dimensions) of the shrinkage of this clay soil upon drying
under laboratory conditions and should provide some indications of the
possible trends under field conditions.

The soil water suction distribution with depth and time when surface
evaporation was prevented is shown for plot 3 in Fig. 4. At the
initiation of the drainage experiment following water ponding (June 15,
1981) it can be seen from the water suction distribution that the soil
was near saturation throughout the soil profile. The water table was
located at approximately 30 cm from the surface where a slight water
suction was exhibited. After nine days of drainage (June 24), the
results indicate the fall of the water table to a depth of approximately
70 cm from surface. However, the water suction near the soil surface
reached only a maximum of 270 cm. As can be seen from the soil water
suction relation (Fig. 3) little change in moisture content occurred
within such a range. The results shown in Fig. 4 represent the slow
drying patterns obtained when a surface cover was maintained to prevent
evaporation. This was due, in part, to the frequent rainfall as well as
the poor drainage characteristics of the Sharkey soil.

Table 1. Dimensions of the soil core samples after the application of
incremental pressures to 1 bar for all three plots of Sharkey clay
soil.***

Soil Layer (cm)	PLOT 1			PLOT 2			PLOT 3		
	L* (cm)	D** (cm)	Volume+ Change (%)	L (cm)	D (cm)	Volume Change (%)	L (cm)	D (cm)	Volume Change (%)
0-17.8	2.8	4.8	26.2	2.6	4.7	34.3	2.8	4.9	23.0
17.8-43.2	2.8	4.9	23.0	2.7	5.0	22.7	2.8	4.8	26.2
43.2-78.7	2.7	5.0	22.7	2.7	4.9	25.8	2.7	4.9	25.8
78.7-106.7	2.6	4.7	34.3	2.6	4.6	37.0	2.7	4.7	31.7
101.7-160	2.5	4.6	39.4	2.5	4.6	39.4	2.6	4.6	37.0

*   Initial length of sample, L = 3 cm
**   Initial diameter of sample, D = 5.4 cm
\+   Initial volume of sample = 68.6 cm^2
*** Average of 3 replications

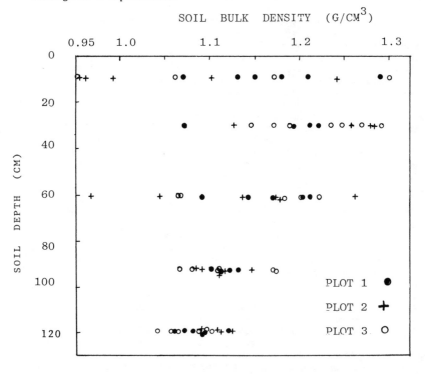

Fig. 1. Soil bulk density with depth for Sharkey clay.

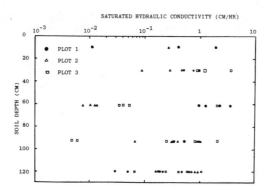

Fig. 2. Laboratory determined saturated hydraulic
conductivity with depth for Sharkey clay.

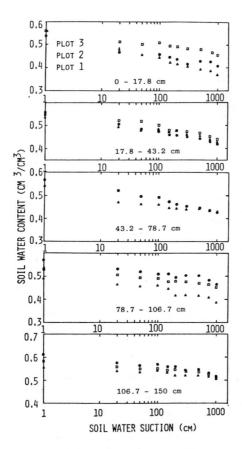

Fig. 3. Soil moisture vs. suction on undisturbed soil samples from five
horizons of Sharkey clay.

In contrast to the soil-water suction distribution under drainage
conditions (Fig. 4), the suction distributions obtained when
simultaneous evaporation and drainage occurred were far more
encouraging. This is illustrated by the drying curves shown in Fig. 5
for plots 1 and 2 for a period of 8 days (May 29 to June 6, 82). During
this period the water table had fallen from approximately 60 cm to 120
cm from the soil surface. In the meantime, the suction at the shallow
soil depths increased drastically with time. As anticipated, the
evaporation method for the determination of in situ hydraulic
conductivity provided a wider range of water contents than did the
drainage method.

A comparison of the soil water content profiles as obtained from
gravimetric sampling with those obtained from the water suctions as
obtained from the tensiometers and the water retention curves are shown
in Fig. 6. The data presented are for May 31 for plots 1 and 2 only.
It is clear that the gravimetrically obtained soil-water contents were
considerably less than those obtained from the indirect method (i.e. the
water retention curves). There is no explanation for this phenomenon
which has been observed for other soils (Scott et al, 1982). It is
possible that laboratory procedures for saturation of the soil (for a
period of 5-7 days) by upward flow allowed complete saturation of the
pores. In contrast, under field conditions of frequent infiltration and
redistribution some pores are bypassed and complete water saturation is
never attained.

In order to calculate the in situ hydraulic conductivity, the potential
water gradients with depth and time must be calculated and the zero plane
of flux must be identified. The plane of zero flux defines the boundary
between upward water flow due to evaporation and that of downward flow
due to drainage. This is essential in calculating the hydraulic
conductivity and it involves the summation of soil moisture content
changes from the plane of zero flux to the desired depth. An example of
the potential and hydraulic gradient distribution at selected times is
shown in Fig. 7 for plot 1. As expected, steeper gradients were
encountered close to the soil surface due to the evaporative flux and
the gradients increased with time. However, no distinct plane of zero
flux was found although the flux values were small at lower depths. The
second disturbing aspect was the presence of upward flow within the
saturated zone, (below the water table). The scatter of the experimental
data (tensiometer measurements) as well as the fact that the results
were not smoothed are possible reasons for this observation. For the
purpose of calculating the in situ hydraulic conductivity, it was found
that the water table can serve as the plane of zero flux as long as only
positive gradients (or upward flow) occur above it. This is simply
because the only changes in the soil moisture content occurred above the
water table when the water content is unsaturated.

Results of in situ unsaturated hydraulic conductivity (K) vs. water
contents are shown in Fig. 8. Data from the top three layers (0 to 78.7
cm) only are presented. No K values were obtained for the lower layers
because little if any changes in the soil water content from saturation
occurred during the drying time considered. The results shown in Fig. 7
show a wide scatter of the K values for all three depths. Moreover no
definite trend of K vs water content can be seen over the range of soil
moisture contents. In general, the range of K values was between 10^{-4} to
10^{-2} cm/hr. Such unsaturated K values were at least one order of
magnitude lower than the laboratory measured saturated K values presented
in Fig. 2.

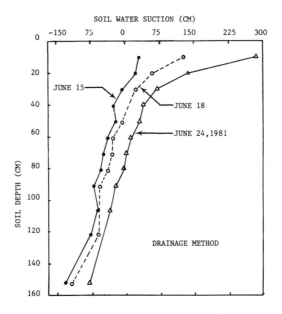

Fig. 4. Soil water suction distributions under
drainage conditions for selected times
during June 15 to June 24, 1981, plot 3.

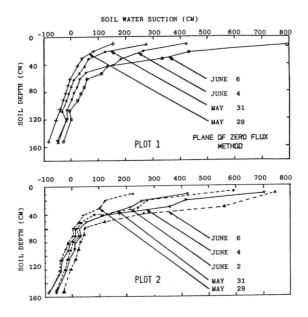

Fig. 5. Soil water suction distributions during
evaporation and drainage (May 29 to June 6,
1982) for plots 1 and 2.

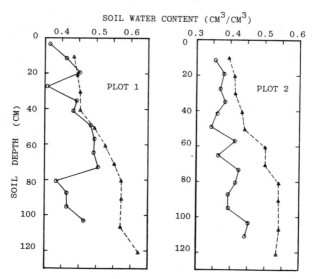

Fig. 6. Comparison of field sampled moisture content
distribution (solid line) with moisture
distribution as determined from the moisture
release curves (dashed line).

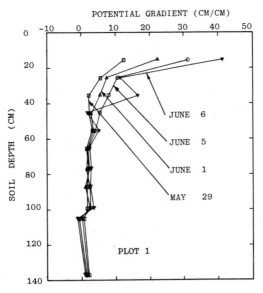

Fig. 7. Potential gradient vs. soil depth for
selected times under evaporation and
drainage conditions, plot 1.

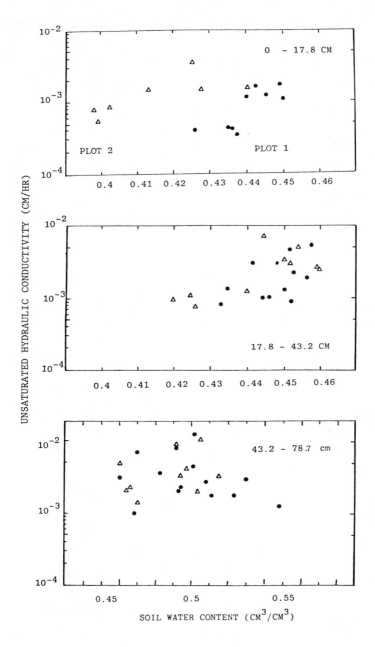

Fig. 8. In situ hydraulic conductivity for the top
three layers of plots 1 and 2 of Sharkey clay.

SUMMARY

In situ soil water behavior was monitored during a twelve-month period
for a Sharkey clay soil in southern Louisiana. This soil is
characterized by its high clay content, extensive shrinkage and swelling,
and the presence of a shallow permanent water table. Three adjacent
plots (2.2 x 2.2 m) were instrumented with three replications of soil
tensiometers at selected depths up to 152 cm. Water suction during water
redistribution was monitored under drainage conditions (i.e. no surface
evaporation) for the purpose of determining in situ hydraulic
conductivity (K) using the instantaneous profile or drainage method.
Such conditions resulted in slow drying patterns over a period of
several months. In contrast, the evaporation or zero plane of flux
method, where water redistribution occured under simultaneous drainage
and evaporation condition proved to be more successful than the drainage
method. Unsaturated K in the top three soil layers ranged from 10^{-4} to
10^{-2} cm/hr in the soil water content range of 0.45 to 0.55 cm^3/cm^3.
Results of bulk density for individual soil layers indicate a high
degree variability in the top soil layers which was probably due to soil
shrinkage and swelling. Soil volume changes or shrinkage which resulted
from drying due to applied suction was quantified also. The results
indicate that all soil layers exhibited some degree of shrinkage
ranging between 22 to 39% of total soil volume.

REFERENCES

1. Arya, L. M., D. A. Farrell, and G. R. Blake. 1975. A field study of
   water depletion patterns in the presence of growing soybean roots.
   I. Determination of hydraulic properties of the soil. Soil Sci.
   Soc. of Am. Proc. 39:424-430.

2. Cassel, D. K. 1975. In situ unsaturated soil hydraulic conductivity
   for selected North Dakota soils. North Dakota State University
   Bulletin No. 494.

3. Klute, A. 1965. Laboratory measurements of hydraulic conductivity
   of saturated soil. In C.A. Black et al. (ed) Methods of Soil
   Analysis. pp. 210-221. Am. Soc. of Agron., Madison, WI.

4. Lund, Z. F. and L. L. Loftin. 1960. Physical characteristics of
   some representative Louisiana soils. ARS 41-33. 83 p.

5. Scott, D. A., R. J. Philips, M. J. M. Romkens, H. M. Selim, and F. D.
   Whisler. 1982. In situ determination of soil hydraulic conductivity
   in a fragipan soil. Agron. Abstracts. Amer. Soc. of Agron.

6. Thomas, G. W. and R. E. Phillips. 1979. Consequences of water
   movement in macropores. J. Environ. Qual. 8:149-152.

7. United States Department of Agriculture. 1968. Soil survey of East
   Baton Rouge Parish. USDA-ARS and La. Agric. Exp. Sta.

8. Watson, K. K. 1966. An instantaneoud profile method for determining
   the hydraulic conductivity of unsaturated porous materials. Water
   Resour. Res. 2:709-715.

9. Wilson, G. V. 1981. Evaluation of different methods for the
   determination of unsaturated soil-water hydraulic conductivity. M.S.
   Thesis, Louisiana State University Library. 134 p.

BACKFILL ALTERATIONS AND DRAINAGE OF CLAY SOILS

George S. Taylor
Assoc. Member ASAE

Norman R. Fausey
Assoc. Member ASAE

On a long term basis, drainage of clay soils is always inadequate. As drainage is improved the intensity of land use increases. There is usually a shift to more shallow rooted annual crops. The average soil moisture contents become greater, and there is more frequent use of heavy equipment. The overall result is deterioration of soil structure, and a greater degree of drainage is required to maintain high crop productivity.

Most cultivated clay soils have low saturated permeabilities. Plowing and incorporation of crop residues usually creates a more permeable topsoil. Shallow cultivation also may increase permeabilities temporarily in the surficial 0.05 to 0.07 m. However, the long term effect of heavy equipment use is soil compaction and reduced permeability below the plow layer. Thus we are often faced with removing excess water from the more permeable plow zone by use of drains installed at depths of low permeability.

This problem and its proposed solutions have been identified in reports from Japan, New Zealand, Belgium, Canada, Israel, England and the United States (Fausey and Hundal, 1980). While improved surface drainage will alleviate the problem, most solutions involve some backfill alteration to enhance drainage of the cultivated layer. Their principal objective is to provide a highly permeable backfill that connects the cultivated layer to the drain, thus permitting water flow to by-pass soil layers of low permeability. Water under positive pressures in the plow layer or interconnected mole channels can readily flow through the highly permeable backfill to the drain.

Information on highly permeable backfill effects on drainage is sparse and sometimes conflicting. Rands (1974) concluded from research on clay soils in England that the use of gravel in the backfill could only be justified where mole channels were made across deeper pipe drains. For economic reasons the upper boundary of the gravel was only a few cm above mole channel depth. He considered the gravel backfill to serve as a hydraulic connector between mole channels and a pipe drain. Armstrong, et al. (1980) also reported an increase in water table drawdown in a sloping clay soil where mole channels were used over pipe drains with a gravel backfill. They also reported that total water removal was approximately the same through the permeable, fall plowed topsoil (without subsurface drains) as with the combination of moles and pipe drains with gravel backfill. Neither Rands nor Armstrong evaluated

Salaries and research support provided by State and Federal Funds appropriated to the Ohio Agricultural Research and Development Center, The Ohio State University, and by the Soil Drainage Research Unit, USDA-ARS. OARDC Journal Article No. 119-82.

The authors are Professor, Department of Agronomy, The Ohio State University; and Soil Scientist, USDA-ARS, Columbus, Ohio respectively.

the effect of a gravel backfill that extended from the drain to the lower boundary of the plow layer. Trafford and Rycroft (1973) reported that pipe drain hydrographs for clay soils in the midlands of England were essentially the same with either gravel or natural soil backfill.

The permeability of natural soil used in trench backfill is thought to depend upon its undisturbed hydraulic conductivity, moisture contents at the time of trenching and backfilling, the type of machinery used for trenching and refilling, and crop and soil cultural practices. Russell (1934) aptly reminds us that a relatively permanent change of structure in the backfill trench soil is mainly a characteristic of clay soils. Levesque and Hamilton (1967) used trench gravel and peat over 10-year old drains in a soil of high clay content in Canada. They reported no significant differences in soil moisture content, water table elevations or grass yields for undisturbed, peat or gravel backfill. Soil conditioners have been studied for their effect on soil aggregate stability for this use (Dierickx and Gabriels, 1976; de Boodt, 1978). Further evaluations are needed under field conditions before they can be recommended for general use.

The authors of this report have conducted drain backfill experiments in a clay soil for the last 4 years (Hundal, et al., 1980; Taylor, et al., 1980 and 1981). These include subsoiling, trenching and refilling with natural soil, and use of gravel. The experiments were conducted on one site under intensive grain cropping and another under continuous vegetative cover. The following is a report of these investigations.

EXPERIMENTAL

The experiments were conducted on two sites near Sandusky, Ohio, both located at the former North Central Branch of the Ohio Agricultural Research and Development Center. The two sites differed only in their management. One has been an ungrazed grass-legume meadow while the other has been in continuous grain production. The experimental sites are located in the glacial Lake Maumee region, and the topography is relatively flat. The climate is humid continental but modified by its close proximity to Lake Erie. Average annual rainfall is only 800-900 mm, but evaporation is quite low during Winter and Spring months. A seasonal, perched water table develops practically every year following snow melt or Spring rainfall, and subsurface drainage is needed for consistently high crop yields.

The soil is a Toledo silty clay and contains approximately 50% clay in the plow layer and 60% in the B horizon. This soil is classified as a poorly drained, fine, illitic, mesic Mollic Haplaquept. The saturated hydraulic conductivity of the plow layer is relatively high compared to that of the subsoil. Median values of 0.44 and 0.06 m/d were reported for the plow layer and subsoil, respectively, by Hoffman and Schwab (1964). Taylor et al. (1970) reported values for comparable depths to be 1.04 and 0.08 m/d. Conductivity values for the plow layer vary widely, depending on tillage and soil compaction. Considerable cracking of the soil occurs during dry periods. A more detailed soils description is reported by Schwab, et al. (1963).

Continuous Meadow Site

This site contains 12 subsurface drains that were installed in 1971 with a floating-beam mole plow. The drains are 50 mm diameter corrugated plastic tubing, 56 m long, installed at an average depth of 0.4 m and an average grade of 0.1%. Drain spacings were 3, 6 and 12 m. A slope of 0.3% exists across the drains, and surface water cannot pond more than 5 to 10 mm due to ground surface roughness. All drains discharge individually into a drainage ditch where flow rates can be measured manually at the outlets. The field was plowed and a corn crop grown in 1972, plowed in 1973 and seeded to wheat with an intercrop seeding of timothy and red clover. Since harvesting the wheat crop in 1974, the field has not been farmed except to clip the clover, timothy, and volunteer grasses twice annually. A nearly complete vegetative cover has existed on the site from 1975 to the present (1982).

Trenching and Refilling Over Drains: The following three backfill treatments with the indicated symbols were established on August 18, 1977:
  --Unaltered Backfill: No alteration in soil overlying the drains since installation in November 1971 (T_0).
  --Trenched and Refilled: Trench excavated directly above and to within 0.05 m of the drain and then backfilled with excavated soil (T_1).
  --Trenched, Spaded and Refilled: Same as T_1 except trench was hand spaded to expose the upper one-third circumference of drain before backfilling (T_2).
For the trenched treatments, a 0.2 m wide trench was excavated with a chain trencher over the entire length of the drain line to an average depth of 0.05 m above the drain. The spaded soil (T_2) was not removed from the trench but left in a rough condition. The trenches were backfilled with the excavated soil, mounded above the drain but not compacted. The soil moisture content during trench excavations was near the lower plastic limit but the soil did not adhere to the trencher blades. The treatments were replicated 3 times, each treatment containing one drain at each of the three spacings.

Subsoiling Over Drains: The following three subsoiling treatments with the indicated symbols were initiated on October 7, 1977 for drains spaced at 12 m:
  --Subsoiled 0.16 m deep (S_1).
  --Subsoiled 0.26 m deep (S_2).
  --Subsoiled 0.34 m deep (S_3).
Subsoiling was done over the entire length of a single drain with a chisel plow having a shank of 0.05 m width. For the S_1 and S_2 treatments, a single run of the plow was made directly over the drain line at the indicated depth. For the S_3 treatment, a first run of the plow was made at about 0.2 m, followed by a second run in the same furrow at a depth of 0.34 m. The soil moisture at the time of subsoiling was near the plastic limit in the subsoil but was drier in the surface layer. Each of the subsoiling treatments brought about appreciable fracturing and loosening of the soil in the upper 0.2 m. The deepest subsoiling (S_3) resulted in the formation of an oblong shaped channel just above the drain. This mole-like channel was approximately 0.07 m wide and 0.07 to 0.15 m high. The slit connecting the channel to the surface soil was partially filled with loose soil. Subsoiling at the intermediate depth (S_2) resulted in the formation of a partial channel in the subsoil. The most shallow subsoiling (S_1) fractured the soil but did not form a channel.

Grain Farming Site

The studies at this site utilized a portion of a long-term experiment that was initiated in 1958 (Schwab et al., 1963). The present study was conducted on two previously established plots, each 37 m by 61 m in dimension and containing 0.22 ha. In 1958 the plots were leveled and an earthen dike of 0.15 m height built around each. In 1969 corrugated plastic tubing of 42 mm diameter was installed within each plot with a floating-beam mole plow to an average depth of 0.4 m, a spacing of 6.1 m, and a slope of 0.17%. There were four drains within the cultivated portion of each plot and two along the plot borders, each being 61 m long. The drains discharged into a manhole where flow rates could be measured. Prior to the backfill study, the site had been grain cropped for 13 of the last 19 years.

For the period of this study, the site was cropped to a corn-soybeans-oats rotation. Each crop was grown annually and occupied one-third of each plot.

The plots were fall plowed each year. Seedbed preparation, planting and crop harvesting was typical of that used by farmers in the area. Plowing and planting were aligned at right angles to the drains.

Backfill Treatments: The backfill treatments were as follows:
  --Unaltered Backfill: No alteration in soil overlying drain since installation in 1969.
  --Gravel Backfill: Trench excavated directly above drain in October 1977. Band of crushed stone placed between plow layer and drain.
  --Trenched and Refilled: Trench excavated directly above drain in October 1978 and refilled with excavated soil.
A light-weight chain trencher was used to excavate a 0.15 m wide trench over the entire length of the drains selected for backfill alteration. A hand spade was then used to expose about 0.6 m lengths of the drain at intervals of 3 to 4 m. About one-third of the drain circumferential area could be seen at that time. A loose layer of wheat straw was placed over the exposed lengths of drains. For the trenched and refilled treatment, a tractor-mounted blade was used to refill the trenches with the excavated soil. For the other trenches, gravel was used to fill the trench to within 0.18 m (approximate plow depth) of the ground surface. Another loose layer of straw was placed over the gravel and the trench was filled and mounded 15-20 cm with the excavated soil. The gravel layer consisted of crushed limestone rock having diameters between 10 and 30 mm. The hydraulic conductivity of the gravel was approximately 4,000 m/d.

The two drains most interior to each plot were selected as control (i.e., "unaltered backfill") because these drains were an integral part of the long-time experiment and their backfill could not be modified. The remaining two drains within the cultivated portion of each plot were chosen at random for the two altered backfill treatments. Flow rates of the drains within each plot were measured periodically for seven years prior to this study. The maximum flow rates measured each year ranged for 0.10 to 0.25 $m^3 day^{-1}$ (m drain)$^{-1}$, the greatest flow rates always occurring from the two drains selected as control.

PROCEDURE

Ponding was created by applying excess water with sprinkler irrigation, once annually at the continuous meadow site and twice annually for the grain farmed plots. Drain flow rates were measured during the irrigations and on some additional dates following rainfall and spring thaw. Priority was placed on obtaining the greatest flow rate during a hydrologic event. Water table elevations were measured by use of perforated pipes installed to a depth of 0.5 m and at the midspacing between drains. The piezometric pressure head inside the drains was measured by use of 15 mm diameter solid wall pipes that were installed about 20 mm inside the drain. Occasionally, drain water samples were collected to determine sediment concentrations. The average percentage ponding of the ground surface was estimated visually following each irrigation.

RESULTS

The data presented herein emphasize the characteristics of drain flow as related to backfill alterations, the longevity of these alterations and water table drawdown. A more complete report can be obtained from articles by Hundal, et al. (1980) and Taylor, et al. (1980 and 1981).

Drain Flow Characteristics

Since flow hydrographs obtained at the two sites are similar, the effects of
backfill treatments on drain flow are illustrated by the hydrographs of the
grain farmed site for a single irrigation in 1979 (Fig. 1, left). This
irrigation was applied the first year after trenching and refilling but the
second year following gravel application. The percentage of the ground
surface covered by ponded water is shown in the lower portion of the graph.

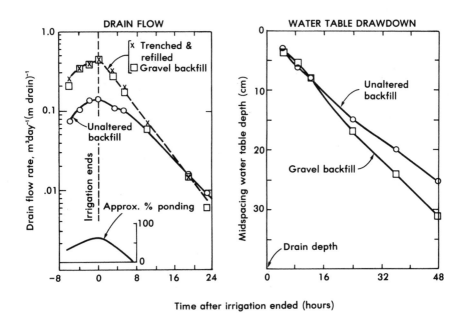

Fig. 1   Flow Hydrographs of Drains with Gravel, Trenched and Refilled,
and Unaltered Backfill; and Water Table Drawdown for Gravel
and Unaltered Backfill. Each Point on the Curves Represents
an Experimental Measurement.

As long as percentage ponding exceeded 20 or 30%, flow rates from drains with
altered backfill were 2 to 4 times greater than those with unaltered backfill.
Flow rates decreased rapidly soon after ponding disappeared, especially for
drains with altered backfill; and thereafter appeared unrelated to the backfill treatment. For both sites, the amount of ponding was always closely
related to drain flow rates.

We were unable to discern whether high water table elevations (without
ponding) had any differential effect on drain flow due to backfill treatment.
The principal difficulty in this evaluation was that the degree of ponding
differed along the length of the drain lines as surface water commenced to
disappear. Thus high water table elevations occurred over some portions of
the drain while ponding occurred over others. The relatively small amounts
of ponding overwhelmed any influence of water table elevation, thus masking
any relationship between drain flow and water table elevation.

Water Table Drawdown

Backfill alterations have not had a significant effect on drawdown, neither
at the continuous meadow nor the grain farmed site. A typical drawdown is
illustrated in the right half of Fig. 1 for the graveled and unaltered backfill at the grain farmed site. The principal reason for nearly identical

drawdown rates for all backfill treatments is that ponding disappeared from over all drains within a 30 min period. This occurred because there was no diking between drain lines, and surface water could move freely across drains with different backfill treatment. If dikes had been installed midway between individual drain lines, it is most likely that drawdown would have been more rapid for drains with greater flow rates. From a practical viewpoint, we may conclude that no improvement in drawdown can be expected from a permeable backfill for sites with good surface drainage.

Maximum Drain Flow Rates

Continuous Meadow Site: The effect of backfill alterations on maximum drain flow rates are shown in Fig. 2 for a 4-year period. Trenching to within 0.05 m of the drain and refilling the trench with soil doubled drain flow rates, while the same treatment plus hand spading essentially tripled flow rates--all compared to drains with no backfill alteration. During the first two years, there was a 20 to 25% reduction in the flow rates of drains that were trenched and spaded. Otherwise, the greater flow rates due to backfill alteration have not declined appreciably for the last four years.

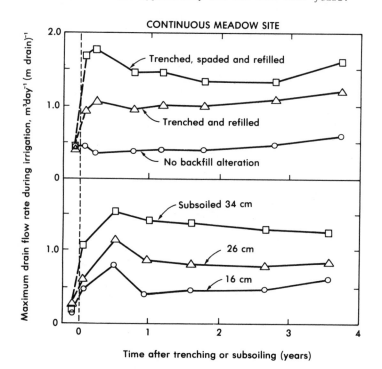

Fig. 2 Effect of Subsoiling or Trenching and Refilling
Directly Over Drains on Flow Rates for Different
Times at the Continuous Meadow Site. Each Point
on the Curves Represents a Flow Measurement Date.

Because there was not a "no backfill alteration" treatment for the subsoiled drains, all comparisons of flow rates are made to those obtained just prior to subsoiling. For some unknown reason, flow rates for these drains prior to subsoiling were quite small. Subsoiling to 0.16, 0.26 and 0.34 m depths, respectively, resulted in stepwise increases in flow rates. Subsoiling to 0.34 m gave flow rates that were approximately the same as trenching and refilling.

Grain Farmed Site: As shown in Fig. 3 both trenching and refilling (upper graph) and gravel (lower graph) increased maximum flow rates. Gravel increased these rates by a factor of 2 to 3. The greater flow rates for these drains have persisted for 4 years; however, there appears to be a small but consistent decline for the last 3 years. For the first year following trenching and refilling, flow rates for these drains were comparable to those from drains with gravel backfill. During the second year, however, flow rates for trenched drains were only 50-60% of those for drains with gravel; and there appeared to be a further decline of some 10% during the third year.

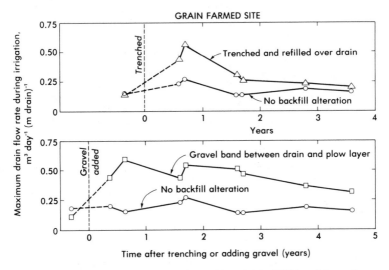

Fig. 3 Effect of Gravel or Trenching and Refilling Directly
        Over Drains on Flow Rates for Different Times at the
        Grain Farmed Site. Each Point on the Curve
        Represents a Flow Measurement Date.

There were variations in annual peak flow rates for drains without backfill alteration at both sites. The variations were less at the continuous meadow than at the grain farmed site. Apparently, the greater variation at the latter site reflects annual differences in tillage and weather conditions. The third-year measurement date for the grain farmed site followed a drought. There was extensive soil cracking, and three pre-irrigations were needed to bring about sufficient reswelling in order to obtain flow rate differences among the treatments.

Sediment in Drainage Water

At the continuous meadow site, sediment concentrations in the drainage water have been quite small and approximately the same for drains that were unaltered, subsoiled or trenched. Sediments concentrations during the period of maximum drain flow were approximately 0.01 kg/L following irrigation but 0.20 kg/L during rainfall. At the grain farmed site, sediment concentrations were also essentially the same for drains that were unaltered, trenched or graveled. However, these concentrations were several times greater than at the continuous meadow site, averaging 0.05 and 0.65 kg/L, respectively, during irrigations and rainfall. For both sites there appeared to be no relationship between drain flow rates and sediment concentrations.

Drain Piezometric Pressures

Positive pressures (i.e. backpressure) were obtained in all drains during ponded flow conditions, apparently due to the small drain diameter. Back-pressures were two-fold greater for drains with altered than with unaltered backfill. In other words, water was transported through the backfill at rates much greater than the drains could transport water to the outlets.

DISCUSSION

A highly permeable backfill between the plow layer and a pipe drain has the potential to increase significantly the rate of water removal by drains when-ever positive water pressures exist within the plow layer. The latter condition can occur whenever a topsoil of moderate permeability is ponded. It can also occur without ponding when heavy rainfall or snow melt occurs on a rough plow layer. A highly permeable plow layer is not uncommon for plowed clay soils before seedbed preparation. Thereafter plow layer permeability will be affected by the amount of tillage, organic matter, clay content and equipment traffic. Once the water table has receded to plow depth, a permeable backfill is not likely to increase the drainage rate.

Sediment concentrations in drainage water were not affected by backfill alterations. As one would expect, greater drain flow due to backfill permea-bility enhancement would increase the total quantity of both flow and sediment load. However, the total sediment load will be much less than if the water was removed through surface runoff (Schwab, et al., 1980).

Longevity of Backfill Alterations

Natural Soil: The effect of backfill alterations on drainage at the two sites differed considerably, depending on whether the land was in meadow or corn. Under continuous meadow, all subsoiling and trenching operations increased drain flow rates under ponded conditions. There has been no significant decline in these greater flow rates for 4 years. While trenching and refilling at the grain farmed site doubled flow rates during the first year, there was a significant decline during the second and third years. It appears that these flow rates may be similar to drains with unaltered backfill within another one to two years. The rapid decline in flow rates was probably a result of clogging of the large soil channels due to loose soil created by tillage and by collapse of these channels due to longer periods of wetness under grain farming.

Although the data are not reported herein, subsoiling was done over four drains at the grain farmed site during a two-year period. There were no increases in flow rates for the year following subsoiling, and no further flow measurements were taken. Excavations were made over the drains immediately after each subsoiling, and only a small amount of soil fracturing was observed. The reason for little fracturing was the relatively high moisture contents that prevailed below plow depth. The opposite situation existed under continuous meadow where the soil was much drier and excellent fracturing occurred.

Gravel Backfill: We can only speculate as to how much longer the drains with gravel backfill will flow at a high rate. The slow decline in these flow rates during the last 3 years suggest a life expectancy of 10-15 years. Three years after installation, a visual examination was made of a 10-m section of gravel. All of the applied straw had disappeared. The gravel had retained some 90% of its original porosity and showed no evidence of clogging. Surprisingly, the gravel that made direct contact with the lower part of the plow layer was relatively free of soil. A mixture of soil and gravel was occasionally found at the junction of the gravel and the drain. The soil appeared to be that left in the trench during gravel installation, rather than soil transported by water from the plow layer.

Practical Implications

A farmer's decision to use permeable materials over pipe drains should be based on the predicted improvements in land drainage and the expected number of years before the backfill materials become clogged with soil. Our data show that a permeable backfill is effective for rapid removal of water under positive pressures. Such conditions frequently exist in low lying areas. Improved drainage of localized wet land areas by use of a permeable backfill may require a relatively small investment. Yet the elimination of a few wet areas within a large field may permit earlier equipment operations in the spring, timely harvesting during wet autumns and an overall increase in crop yield.

The experimental data reported here are expected to be most applicable to soils having a permeable plow layer. The results would apply to many clay soils but probably not to soils high in silts and fine sands. They probably would not apply to "no till" farming unless some fracturing of the plow layer could be done. Since a plow layer of relatively high permeability is needed for a hydraulic connector to be effective, excessive soil compaction and tillage would negate many of the results reported in this study.

REFERENCES

1. Armstrong, A.C., C. Carter, and K. Shaw. 1980. The hydrology and water quality of a drained clay catchment - A preliminary report on the Cockle Park drainage experiment. MAFF, Land Drainage Service, Research and Development Report No. 3, 15 pages. England.

2. deBoodt, M.F.L.P. 1978. Artificial soil stabilization of backfill in field drainage. J. Sci. of Food and Agr. 29(5):417-422.

3. Dierickx, W., and D. Gabriels. 1976. Stabilizing backfill of drain pipes and drainage efficiency. Med. Fac. Landbowuu. Rijks Univ. Gent. 41(1):293-300.

4. Fausey, N.R., and S.S. Hundal. 1980. Role of trench backfill in sub-surface drainage--a review. TRANSACTIONS of the ASAE 23:1197-1200.

5. Hoffman, G.J., and G.O. Schwab. 1964. Tile spacing prediction based on drain flow. TRANSACTIONS of the ASAE 7:444-447.

6. Hundal, S.S., G.S. Taylor, and N.R. Fausey. 1980. Backfill effects on subsurface drainage of Toledo soils: I. With grass-legume cover. Ohio Agric. Res. Dev. Cent., Res. Cir. 257.

7. Levesque, M., and H.A. Hamilton. 1967. The efficiency of a subsurface drainage system with porous filling material in a Kapuskasing clay soil. Canad. J. Soil Sci. 47:55-65.

8. Rands, J.G. 1974. The value of permeable backfill (PF) based on field experiments. In Proceedings of the conferences at Stoneleigh and Church Stretton On the Use of Permeable Backfill. FDEU. Tech. Bull. No. 74/11- England.

9. Russell, J.L. 1934. Scientific research in soil drainage. J. Agr. Sci. 24:544-573.

10. Schwab, G.O., N. R. Fausey and D.E. Kopcak. 1980. Sediment and chemical content of agricultural drainage water. TRANSACTIONS of the ASAE 23:1446-1449.

11. Schwab, G.O., T.J. Thiel, G.S. Taylor, and J.L. Fouss. 1963. Tile and surface drainage of clay soils. Ohio Agric. Res. Dev. Cent., Res. Bull. 935.

12. Taylor, G.S., E. Stibbe, T.J. Thiel, and J.H. Jones. 1970. Hydraulic conductivity profiles of Toledo and Miami soils as measured by field monoliths. Ohio Agric. Res. Dev. Cent., Res. Bull. 1025.

13. Taylor, G.S., S.S. Hundal, and N.R. Fausey. 1980. Permeable backfill effects on water removal by subsurface drains in clay soil. TRANSACTIONS of the ASAE 23:104-108.

14. Taylor G.S., S.S. Hundal, N.R. Fausey, and G.O. Schwab. 1981. Backfill alteration effects on pipe drainage of a clay soil. SSSAJ 45(3):611-616.

15. Trafford, B.D., and D.W. Rycroft. 1973. Observations on the soil water regimes in a drained clay soil. J. Soil Sci. 24(3):380-391.

"SUBSURFACE DRAINAGE: A COMPARISON BETWEEN FIELD EVALUATION AND ESTIMATED

PROPERTIES FOR FOURTEEN SOILS IN THE MIDWEST WITH RESTRICTIVE LAYERS"

by Paul E. Lucas[1]
Member ASAE

Drainage of slowly permeable clayey soils having a very slowly permeable restrictive layer has always presented unique management problems. In the past, general drainage recommendations for most of these soils in the Midwest have been for surface drainage methods only. During the seventies, increased subsurface drainage activity was noted on these soils. The underlying motive for subsurface drainage installed in these soils could be attributed to (1) increased usage of plastic tubing, (2) heavier machinery requiring drier footing, (3) larger land units and capital investment in large equipment making plant-harvest periods more critical, (4) more effective control of perched water tables for optimum crop root development, and (5) improved crop varieties which often require better drainage to obtain optimum crop yields.

The objective of this paper is to compare subsurface drainage field evaluations on 14 heavy-textured soils with recommendations calculated from drainage formulas using estimated soil properties.

FIELD EVALUATIONS

Because of a need for more reliable information of the subsurface drainage performance of these soils, the Soil Conservation Service has contracted with Ohio State University and the University of Illinois to perform field evaluations. Evaluations were made on the following soils:

Soil   Classification*

Avonburg silt loam--fine-silty, mixed, mesic Aeric Fragiqualfs
Bonnie silt loam--fine-silty, mixed, acid, mesic Typic Fluvaquents
Bryce silty clay loam--fine, mixed, mesic Typic Haplaquolls
Clermont silt loam--fine-silty, mixed, mesic Typic Ochraqualfs
Darwin silty clay--fine, montmorillonitic, mesic Vertic Haplaquolls
Hoytville silty clay loam--fine, illitic, mesic Mollic Ochraqualfs
Latty clay--fine, illitic, nonacid, mesic Typic Haplaquepts
Nappanee silty clay loam--fine, illitic, mesic Aeric Ochraqualfs
Paulding clay--very fine, illitic, nonacid, mesic Typic Haplaquepts
Piopolis silty clay loam--fine-silty, mixed, acid, mesic Typic Fluvaquents
Roselms clay--very fine, illitic, mesic Aeric Ochraqualfs
Toledo silty clay--fine, illitic, nonacid, mesic Mollic haplaquepts

*Soil Taxonomy (Soil Survey Staff, 1975)

Procedures used in the field evaluations are discussed by Patronsky and Schwab (1979) and in Ohio Research Bulletin 1141 (Schwab et al., 1982). The effective values of hydraulic conductivity and drainable porosity were obtained by procedures given by Skaggs (1976), who modified those developed by Hoffman and Schwab (1964). Skaggs gave procedures and calculations by which

---

The author is: Paul E. Lucas, Drainage Engineer, Midwest National Technical Center, Soil Conservation Service, Lincoln, Nebraska.

the drainable porosity could be estimated for a soil by measuring the time-associated drawdown curves for an in situ subsurface drainage system.

The drainable porosity values determined from the evaluations are given in Table 3 of the reports (Schwab et al., 1978-79; Walker et al., 1981). The weighted values were given for the 0-10, 0-20, and 0-30 cm depths. Following Table 1 is a typical Table 3 of one of the "Drainability" reports of Schwab et al. (1979). All soils evaluated were located in Ohio unless otherwise indicated.

Table 1. Average Measured Drainage Porosities by Soil Series

Soil Type	Plot No.	Test No.	Initial Wt. Avg. Drawdown (cm)	Drainage Porosities by Depth Increments		
				0-10	0-20	0-30
				cm	cm	cm
Hoytville	1	1	2.7	0.0532*	0.0471	0.0386
		2	1.3	.0657	.0479	.0446
	2	1	2.6	.0758	.0675	.0535
		2	4.3	.0734	.0639	.0524
	Average		2.7	.0670	.0566	.0473
Clermont (Indiana)	1	1	0.9	.0749	.0576	.0484
		2	2.3	.0417	.0394	.0392
	2	1	1.1	.0503	.0492	.0438
		2	1.5	.0773	.0568	.0491
	Average		1.5	.0611	.0508	.0451
Paulding	1	1	2.2	.0817	.0706	.0672
		2	1.4	.1246	.0864	.0890
	2	1	6.1	.0959	.0816	.0693
		2	3.9	.1083	.0642	.0654
	Average		3.4	.1026	.0757	.0727
Nappanee	1	1	4.1	.0721	.0696	.0535
		2	3.0	.0793	.0646	----
	2	1	5.1	.0688	.0674	.0546
		2	3.5	.0958	.0692	----
	Average		3.9	.0790	.0677	.0541

$*cm^3/cm^3$

For comparative analyses, the drainable porosities for a given drawdown test were segregated to the 0-10, 10-20, and 20 plus cm horizons by calculation from the weighted values.

The drawdown curves for two different time periods were plotted from the evaluation data (Fig. 1). The area between the drawdowns were planimetered for the 0-10, 10-20, and 20 plus cm horizons. These values were then multiplied by the horizon segregated drainable porosity values (ϕ_d) and then divided by the tile spacing (L) and the difference in time values (ΔT) to obtain the drainage coefficient values (D.C.$_{\Delta T}$). The D.C.$_{\Delta T}$ values were multiplied by 24 hours to give D.C.$_{24}$ values in mm/d.

The equation for the calculations is:

$$D.C._{24} = \frac{A_{0-10} \times {}^{\phi}d_{0-10} + A_{10-20} \times {}^{\phi}d_{10-20} + A_{20\ plus} \times {}^{\phi}d_{20\ plus}}{L \times \Delta T} \times 24 \times 10$$

where: A_{0-10} = drawdown area between 0 and 10 cm, square cm

A_{10-20} = drawdown area between 10 and 20 cm, square cm

$A_{20\ plus}$ = drawdown area greater than 20 cm, square cm

${}^{\phi}d_{0-10}$ = drainable porosity between 0 and 10 cm, cm/cm

${}^{\phi}d_{10-20}$ = drainable porosity between 10 and 20 cm, cm/cm

${}^{\phi}d_{20\ plus}$ = drainable porosity for depths exceeding 20 cm, cm/cm

L = drain spacing, cm

ΔT = time difference for drawdown to occur, hours

$D.C._{24}$ = drainage coefficient, mm/d

Figure 1 shows a typical plot for the Nappanee soil.

The drainage coefficient values obtained by this method will be used in the comparative analysis for another method to obtain drain depth and spacing.

Whereas the field evaluations were used to establish actual performance, the question can be asked whether performance could have been estimated from other sources.

SOIL INTERPRETATIONS RECORD (SIR)

The Soil Conservation Service estimates the permeability range by major layer for soil series as a part of the data provided for the Cooperative Soil Survey. The layers are identified mainly by texture and to a lesser degree, other properties that can change appreciably even for a soil series with all the layers having the same texture. Estimates are normally given to a depth of 1.5 m or to bedrock if less than 1.5 m. Figure 2 shows a portion of the data given in an SIR.

Permeability is estimated by the procedures given by Uhland and O'Neal (1951) which use the soil properties--mainly texture, structure, sodium, sodium content and the bulk density. Other methods to supplement these estimates are field tests using the Core Extracting Method (Black et al., 1965) for horizons above the water table and the Auger-Hole Method for layers below the water table.

There is no attempt to correlate the permeability rates to the lateral hydraulic conductivity which is the normal parameter to determine subsurface drainage performance. Permeability is generally shown on the SIR in the following ranges-- < 1.5 mm/hr; 1.5 to 5.1 mm/hr; 5.1-15.2 mm/hr; 15.2-50.8 mm/hr; 50.8 to 152.4 mm/hr; 152.4 to 508.0 mm/hr; and > 508.0 mm/hr.

By use of the permeability ranges shown in the SIR's and the Hooghoudt Drainage Equation (ILRI, 1972), the drain depth-spacing relationship can be established. This is used as the means to compare drainage performance by estimated data to that obtained by field evaluation.

OHIO
NAPPANEE SOIL PLOT I, TEST I
CROSS SECTION A

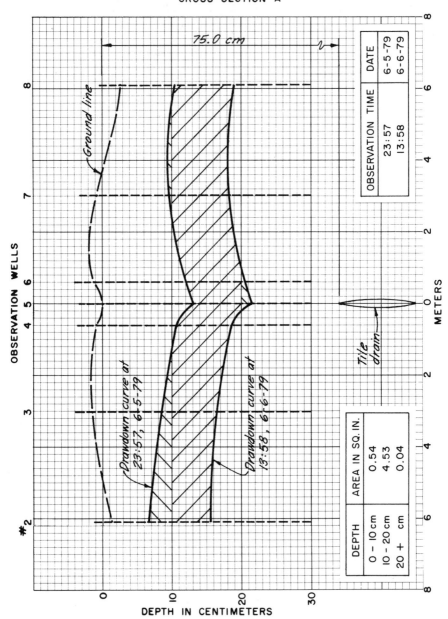

Fig. 1   Drawdown performance from field evaluation.

MLRA(S): 108, 110   BRYCE SERIES
REV. LMK, 9-81
TYPIC HAPLAQUOLLS, FINE, MIXED, MESIC

THE BRYCE SERIES CONSISTS OF POORLY DRAINED SOILS FORMED IN CLAYEY SEDIMENTS AND UNDERLYING CLAYEY GLACIAL TILL ON
MORAINES AND TILL PLAINS. THE SURFACE LAYER IS BLACK SILTY CLAY 13 INCHES THICK. THE SUBSOIL IS BLACK, DARK GRAYISH
BROWN, OLIVE GRAY AND GRAY MOTTLED SILTY CLAY IN UPPER 32 INCHES AND GRAY MOTTLED SILTY CLAY IN LOWER 15 INCHES. SLOPES
ARE 0 TO 2 PERCENT. CROPLAND IS THE MAIN USE.

ESTIMATED SOIL PROPERTIES

DEPTH (IN.)	USDA TEXTURE	UNIFIED	AASHTO	FRACT >3 IN (PCT)	PERCENT OF MATERIAL LESS THAN 3" PASSING SIEVE NO. 4				LIQUID LIMIT	PLAS- TICITY INDEX
					4	10	40	200		
0-13	SIC, SICL	CH, CL	A-7	0	100	100	95-100	85-95	45-60	20-30
13-45	SIC, C	CH	A-7	0-5	95-100	95-100	95-100	75-95	50-60	25-35
45-60	SIC, SICL, C	CH, CL	A-7	0-5	95-100	90-100	90-100	75-95	40-65	20-40

DEPTH (IN.)	CLAY (PCT)	MOIST BULK DENSITY (G/CM3)	PERMEA- BILITY (IN/HR)	AVAILABLE WATER CAPACITY (IN/IN)	SOIL REACTION (PH)	SALINITY (MMHOS/CM)	SHRINK- SWELL POTENTIAL	EROSION FACTORS		WIND EROD. GROUP	ORGANIC MATTER (PCT)	CORROSIVITY	
								K	T			STEEL	CONCRETE
0-13	35-45	1.30-1.50	0.2-0.6	0.12-0.21	5.6-7.8	−	HIGH	.28	5	4	5-7	HIGH	LOW
13-45	42-52	1.35-1.60	0.06-0.2	0.09-0.13	6.6-8.4	−	HIGH	.28					
45-60	38-60	1.60-1.75	<0.2	0.03-0.05	7.4-8.4	−	HIGH	.28					

FLOODING			HIGH WATER TABLE			CEMENTED PAN		BEDROCK		SUBSIDENCE		HYD POTENT'L
FREQUENCY	DURATION	MONTHS	DEPTH (FT)	KIND	MONTHS	DEPTH (IN)	HARDNESS	DEPTH (IN)	HARDNESS	INIT. (IN)	TOTAL (IN)	GRP	FROST ACTION
NONE			+1-1.0	APPARENT	FEB-JUN	−		>60		−		0	HIGH

Fig. 2.   Selected Data from a Soil Interpretations Record.

Table 2 shows the spacing obtained by use of both the low and the high permeability values for the range shown on the SIR. In those situations where the permeability is shown as < 1.5 mm/hr, that value was used for <u>both</u> the low and high value since there was no apparent way to quantify a <u>lower</u> limit.

Drawdown values for the Hooghoudt steady-state drainage equation were set at the average high values from the drawdown plots. Depth to the actual barrier (permeability = 0) were not shown on the SIR's for the soils evaluated; an arbitrary value of 3.05 m was used in the calculations. Drain size was used as was noted in the field evaluation reports. All were 101.6 mm diameter, except for the Bryce where the drains were 127 mm.

The drain depth and drainage coefficient values shown in columns 2 and 3 in Table 2 were used with the Hooghoudt Equation and the SIR permeability values to obtain an SIR-based spacing. These values are shown in column 5 for the low SIR permeability values and in column 6 for the high SIR permeability values.

Those soils (four of the fourteen) where the spacing range calculated from SIR permeability estimates spanned the actual field spacing are indicated by the *. Soils (eight of the fourteen) where the spacing range calculated from SIR permeability estimates were less than the actual field spacing are indicated by the †. Two soils had spacing ranges calculated from SIR permeability estimates greater than the actual field spacing and are indicated by the #.

Table 2. Comparison of Field Performance vs. Estimates Using SIR Data.

(1)	(2)	(3)	(4)	(5)	(6)
	Drain	Drain	Drain	Equiv. SIR Spacing, m	
	Depth	Coefficient	Spacing,	Low	High
	$m^{1/}$	$mm/d^{2/}$	$m^{1/}$	Perm.	Perm.
Toledo silty clay	0.94	7.0	12.2*	7.6	15.2
Roselms clay	1.1	4.9	10.1#	10.4	12.5
Paulding clay	0.73	9.0	10.7†	4.9	5.2
Latty clay	0.66	5.7	13.1†	6.4	9.1
Clermont silt loam (OH)	0.96	5.4	15.0†	9.1	11.3
Avonburg silt loam	0.9	3.5	15.6#	16.2	25.0
Hoytville silty clay loam	0.96	8.4	12.1*	10.1	20.1
Clermont silt loam (IN)	0.97	7.6	15.2†	8.2	10.4
Nappanee silty clay loam	0.82	7.5	12.3†	6.4	6.7
Bryce silty clay	0.82	9.0	19.7†	10.4	12.2
Bonnie silt loam	0.92	4.6	15.7*	11.6	65.2
Wabash silty clay	0.68	3.5	15.1†	8.2	8.2
Darwin silty clay	0.85	3.8	14.0*	10.4	16.8
Piopolis silty clay loam	0.8	6.0	15.4†	7.0	14.6

$\underline{1/}$ As situated in the field evaluation plots.

$\underline{2/}$ As derived from the field evaluations.

* Field spacing within that calculated for the SIR permeability range.

† Field spacing is greater than that calculated for the SIR permeability range.

Field spacing is less than that calculated for the SIR permeability range.

SUMMARY

A reasonable value of drain spacing generally appears to be obtainable by the Hooghoudt Equation and the estimated soil permeability values as shown on the SCS soil interpretation records. If the SIR values would have been used for four of the fourteen soils, subsurface drainage design estimates would have been acceptable. Estimates of drain spacing for eight soils would have been overdesigned (spacing too narrow) and would have been underdesigned (spacing too wide) for two soils.

A refined drainage performance value is obtained by field performance "drainability" studies. Only by such studies can it be determined whether the actual performance of a soil is at the low or the high end of (or above or below) values estimated by using drainage equations and estimated permeability ranges.

ACKNOWLEDGEMENTS

The author wishes to recognize the contributions of W. J. Ochs and R. J. Patronsky for the technical aspects of this paper and to D. J. Rejda for extracting the data from the "drainability study" computer printouts and his drafting work.

"REFERENCES"

1. Black, C. A. et al., 1965. Methods of Soil Analysis. Agronomy Monograph No. 9, Am. Soc. of Agronomy.

2. Hoffman, G. J. and G. O. Schwab. 1964. Tile spacing prediction based on drain outflow. Trans. of the ASAE, 7(4):444-447.

3. International Institute for Land Reclamation and Improvement (ILRI). 1972. Velboek Voor, Land-en, Waterdeskundigen. The Institute, Wageningen, The Netherlands.

4. Patronsky, R. J. and G. O. Schwab. 1979. A plan for field evaluation of subsurface drains in heavy soils and soils with a restrictive layer. Eighth Technical Conference, U. S. Committee on Irrigation Drainage and Flood Control. International Commission on Irrigation and Drainage. September 26-29, 1979, Phoenix, Arizona.

5. Schwab, G. O. et al., 1978. Phase I research report on field evaluation of subsurface drainage. Contract No. AG-31SCS-01771. OARDC Proj. No. SS-266.

6. Schwab, G. O. et al., 1979. Phase II research report on field evaluation of subsurface drainage. Contract No. AG-31SCS-01771. OARDC Proj. No. SS-266.

7. Schwab, G. O., D. W. Michener, and D. E. Kopcak. 1982. Spacing for subsurface drains in heavy soils. Ohio Agricultural Research and Development Center--Research Bulletin 1141.

8. Skaggs, R. W. 1976. Determination of the Hydraulic Conductivity-Drainable Prosoity Ratio from Water Table Measurements. Trans. of the ASAE, 19(1):73-80.

9. Soil Survey Staff. 1975. Soil Taxonomy: A Basic System of Soil Classification for Making and Interpreting Soil Surveys. USDA Handbk. No. 436. U.S. Government Printing Office, Washington, D.C.

10. Uhland, R. E. and A. M. O'Neal. 1951. SCS-TP-101 USDA.

11. Walker, P. N. et al., 1981. Field evaluation of subsurface drainage. Contract No. 58-6526-9-241.

SUBSURFACE DRAINAGE OF A HEAVY ALLUVIAL SOIL

H. Kuntze   F.A. Wetjen

Soils with low permeability need a narrow spacing of subsurface drains. Even with a narrow spacing, the benefits of drainage often are limited to the area nearest to the drain. Also, costs increase with decreasing drain spacing.

Because of the above, the question arose whether a systematic pattern of subsoiling perpendicular to drains which are widely spaced could bring better results of reclamation (Schulte-Karring 1979). However, the heavy soils in a humid climate as in Northwest Europe do not dry frequently and deeply enough. Consequently, subsoiling is ineffective and frequently damages the soil structure (Renger, et al., 1974). For heavy soils ($>$ 30%$<$2µm, plastic range $>$9) mole drainage seems to be an alternative (DIN 1185, 1973). In order to determine whether mole drainage would be effective, the following questions had to be answered:

1.) How long should a mole drain be?

2.) How should the tile drains be backfilled to get an optimal connection with the mole drains?

3.) How often should the mole drainage be repeated?

SOIL

Most alluvial soils have a profile with a claypan over valley sand and gravel. Our experiments (Wetjen, 1982) were conducted on an alluvial soil of the Weser valley which has a 150-200 cm claypan. The topsoil (-40 cm) is a silty loam, the subsoil consists of silty clay loam and the deeper parts of silty clay. Depending on the changing soil moisture the soil structure of the upper parts of the profile is polyedric, followed by prismatic and coherent elements. The bulk density ranges from 1.6 to 1.4 g/cm^3. Only in the topsoil are enough draining pores $>$ 10 µm to make sufficient aeration possible. Holes of the earthworm are the only big pores in the subsoil. Therefore, the permeability is rather low (1-3 cm/day (Table 1 and Figure 1).

The authors are: Dr.H.KUNTZE, Professor, & Dr.F.WETJEN, Dipl.Ing. Soil Technological Institute, Bremen, FRG.

Table 1. Characteristics of an Alluvial Soil

Depth (cm)	Horizon	Texture, Color and Structure	
0 - 28	Ap	light humic silty loam, Subpolyeders	5 YR 4/3
- 57	Go	silty loam, Polyeders, 2-6 mm	7,5 YR 4/3
- 85	Go/r	silty clay loam Polyeders, 2-8 mm iron dashes	10 YR 5/2
- 140	GR/o	silty clay prismatic structure 10% iron concretes Ø 2-3 mm	5 YR 5/1
- 200	Gr	silty clay coherent, 20% iron concretes Ø 3-5 mm 3% black, manganese concretes	5 YR 5/2

FIELD EXPERIMENTS

In 1975, a relativ dry year, a 4 ha experimental plot was laid
out. The four treatments were: a control without drainage and 3
parts with subsurface collector drains spaced at 16 m, 32 m and
64 m. The average depth of subsurface collector drains was 1 m.
Corrugated PVC-pipe, 65-80 mm in diameter fully wrapped with cocosfilter was used.

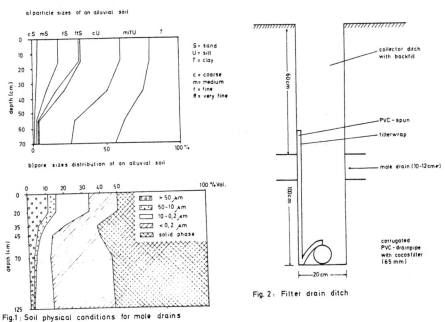

a) particle sizes of an alluvial soil

S = sand
U = silt
T = clay

c = coarse
m = medium
f = fine
ff = very fine

b) pore sizes distribution of an alluvial soil

> 50 μm
50-10 μm
10-0,2 μm
< 0,2 μm
solid phase

Fig.1: Soil physical conditions for mole drains

collector ditch with backfill

PVC - spun
filterwrap

mole drain (10-12 cmø)

corrugated PVC - drainpipe with cocosfilter (65 mm)

Fig. 2: Filter drain ditch

Each of the blocks with collector drains had three backfill treatments:
a) dry soil material out of the ditch
b) gravel (2 - 16 mm) 0.1 m^3/m, and
c) PVC-spun yarn 2 x 50 cm, wrapped with a Duo-filter (acryl + cellulose) (Fig.2).

After backfilling the collector ditches, mole drains at a spacing of 2 m and a depth of 50 - 70 cm were installed perpendicular to the tile drains.

Continuous recording equipment was used as much as possible for the field observations. This included groundwater level by observation wells, discharge of drains by drainflowmeters (Trafford, 1975), and soil moisture by tensiometers.

After 3 and 5 years bulk density, pore space distributions, shear resistance and stability of aggregates (Burghardt, 1982, Wetjen, 1982) were measured to determine any changes of soil structure at different depths and locations. Only small grains were grown during the investigation period. The yields were measured by small split plots (25 m^2)with 4 replications.

RESULTS

Trafficability

It is known that this soil is only trafficable if the groundwater level lies deeper than 75 cm below the soil surface. Higher groundwater levels decrease the bearing capacity and increase the potential of damage to the soil structure with subsequent losses of yield if heavy machines are driven on the soil.

Alluvial soils, Aquepis, with a small amount of large pores like the soil used in these experiments have a large groundwater amplitude. In summer the groundwater level drops to >2 m below soil surface. In the winter months it rises into the topsoil if there is no artificial drainage. Depending on the weather conditions, these alluvial soils can be too wet for agricultural operations in autumn and spring. Tilling will be done too late in the spring and can be hindered in the autumn. Figure 3 shows the influence of the different drainage densities on the available field working days. The periods with a groundwater level higher than 75 cm below the soil surface are shorter if the spacing of the collector drains is less than 32 m. A 6 year average shows that the plots with the 16 m collector drain spacing had only 63 days with groundwater levels higher than 75 cm below the soil surface. With increasing collector drain spacing wet days increase to 85 days at a 32 m spacing and 119 days at a 64 m spacing. Without any artificial drainage the groundwater level was too high for 114 days. The gain in field working days is 51 and 29 days for the 16 m and 32 m collector drain spacing, respectively.

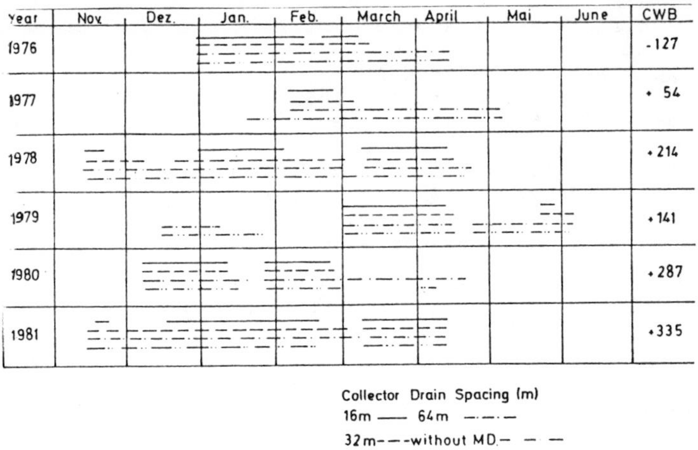

Year	Nov.	Dez.	Jan.	Feb.	March	April	Mai	June	CWB
1976									-127
1977									+ 54
1978									+214
1979									+141
1980									+287
1981									+335

Collector Drain Spacing (m)
16m ——— 64m —·—·—
32m———without M.D.— — —

Figure 3: Groundwater level higher than 75cm below soil surface and climatical
water balance (CWB)

Discharge

From 1977 to 1981, eleven discharge periods were recorded between November and April. On the average one discharge period
continued 28 days at a collector drain spacing of 16 m. With
increasing collector drain spacing the discharge periods continued 31 days and 44 days at the 32 m and 64 m spacings,
respectively (Table 2).

Table 2. Combined Drainage Discharge

Plot	Days of Discharge	Discharge % of rain-fall	Daily Discharge (mm/day)	Highest Discharge (l/s · ha)
16 m	28.4	64	1.05	
+ Gravel		58		2.5
+ Soil backfill		44		2.4
+ PVC		52		4.0
32 m	31.3	41	0.75	
+ Gravel		52		0.9
+ Soil backfill		31		1.2
+ PVC		14		0.3
64 m	44.3	23	0.43	
+ Gravel		23		0.3
+ Soil backfill		12		0,5
+ PVC		22		1,2
without mole drainage	40.7	25	0.41	0.4

The shorter the time of discharge the higher the discharge intensity. During the period of the experiment 755 mm rainfall was
measured. The percentage of discharge was 64, 41 and 23 % for
the 16 m, 32 m, and 64 m collector drain spacings, respectively.
Without artificial drainage 25 % of rainfall was discharged.

The backfill with gravel and PVC-spun-yarn take the water from
the mole drains somewhat faster than only a soil backfill. The
highest daily discharge values show these differences less well.

Pumping stations for artificial drainage normally are designed
for a daily discharge of 1 l/s/ha. Therefore, if mole drainage
is used extensively a pumping station can be undersized for
short periods of time because discharges can be as high as
4 l/s·na.

Up to the present, there has been no correlation between the
time after mole drain installation and discharge. This means
that mole drains in a heavy silty clay loam can work well for at
least 7 years. Three and 5 years after installation the mole
drain itself could not be found, but there was a very loose soil
structure with a high permeability about 20 cm on both sides of
the original mole drain (Burghardt, 1982).

Yields

The average grain yields in dt dry matter (DM) per hectare of
the experimental field depended on the kind of plant and climatical water balance (Table 3).

Table 3. Grain Yield (dt/ha) and Climatical Water Balance (CWB)

===
Year	Crop	Yield	CWB
		(dt/ha)	(mm)
1976	winter wheat	54.2	− 127
1977	winter barley	47.1	54
1978	winter barley	41.9	214
1979	summer wheat	44.3	141
1980	winter barley	57.1	287
1981	winter barley	46.2	335
	6 year average	48.2	
LD 5 %		1.5	
LD 1 %		2.0	
===

Table 4 shows the relationship between yields, collector drain
spacing and backfill material for a 6 year average.

Table 4. Yield Versus Collector Drain Spacing Versus Backfill
Material

===

Collector Drain Spacing (m)	Yield (dt/ha)	Backfill Material	Yield (dt/ha)
16	49.5	gravel	48.4
32	50.0	PVC	48.8
64	45.9	soil	48.2
without mole drainage	48.3		
LD 5 %	1.1		1.0
LD 1 %	1.1		1.4

===

The 16 m and 32 m collector drain spacing gave significant yield
differences from the 64 cm collector drain spacing and no artificial drainage. There was no difference in yield due to the different backfill material.

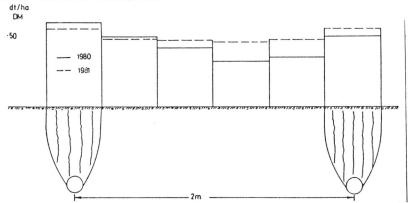

Fig.4: YIELDS OF GRAINS
depending on the distance of mole drains
—alluvial SCL—

A total yield increase of 4 % is rather low. However, if the
yield was harvested in 40 cm strips parallel to the mold drains
a 16 % increase in yield was found directly above the mole
drain (Figure 4).

CONCLUSIONS

The following conclusions can be drawn:

1) The closer the collector drain spacing, the longer the
groundwater level remains deep enough for arable conditions.
The collector drains spaced closer together have a higher rate
of discharge. Both lowering of the groundwater and an increase

in the rate of discharge increase grain yield. A collector drain spacing of 16 m or 32 m with mole drains spaced 2 m perpendicular to the collector drains is suitable for drainage of a heavy alluvial soil.

2) There was no significant differences on the rate of discharge between backfill materials of dry soil, gravel and a special PVC-spun-yarn.

3) Even if the mole drain disappears within a few years the loose structure around its installation keeps the permeability high and supports the drainage of the field.

REFERENCES

1) Burghardt, W.a.Ch.Pätzold, 1982: Soil physical conditions around a collapsed mole drain. Soil Science, in press.

2) DIN 1185, 1973: Dränung-Regelung des Bodenwasserhaushaltes durch Rohrdränung, Rohrlose Dränung und Unterbodenmelioration Blatt 1-5, Beuth-Verlag, Berlin-Köln.

3) Renger, M., et al., 1974: Beurteilung bodenkundlicher, kulturtechnischer und hydrologischer Fragen mit Hilfe von klimatischer Wasserbilanz und bodenphysikalischen Kennwerten. 3.Bericht: Tiefenbearbeitbarkeit Z.f.Kulturtechn.u.Flurberein.15, 263-271.

4) Schulte-Karring, H., 1979: Das Ahrweiler-Meliorationsverfahren zur Beseitigung von Bodenverdichtungen. 8th Conference of the International Soil Tillage Research Organization ISTRO, Stuttgart-Hohenheim.

5) Trafford, B.D., 1975: Improving the design of partical field drainage. Soil Science 199, 334-338.

6) Wetjen, F.A., 1982: Kombinierte Dränung auf einem Auen-Gley, Dr.Thesis, Göttingen.

CROP GROWTH AFTER TRANSIENT WATERLOGGING

R.Q. Cannell   R.K. Belford

The design of field drainage systems is often empirical, being based on
local practice (e.g. Raadsma 1974), such that the control of the water-table
may not be optimal for the crop in relation to the investment being made.
The lack of knowledge of the response of crops to high water-tables at
different stages of growth has been identified by Bouwer (1974) and Trafford
(1975) as limiting progress towards more rational drainage design, an aspect
emphasized by Hiler (1976) at the last National Drainage Conference. We
began an experimental program in 1973 to investigate the effects of
transient waterlogging on the growth and yield of the main crops grown in
the United Kingdom, using lysimeters containing soil monoliths, with
facilities for control of the depth of the water-table and of rainfall. It
was decided not to do this work in the field because of the inherent
variability of the large areas of land that are needed and because of the
uncontrolled effect of rainfall on water-tables. The purpose of this paper
is to summarize the main findings in this program. Detailed results have
been published elsewhere (Belford et al. 1980; Cannell et al. 1980b;
Cannell and Belford 1980; Cannell et al. 1981a,b; 1982).

THE LYSIMETER SYSTEM AND TREATMENTS

To simulate field conditions the experimental system consisted of lysimeters
containing undisturbed soil monoliths, supported in a suitable structure
buried in the surrounding soil, with access provided below ground for the
control of water-table height and collection of drainage water. A movable
glasshouse excluded unwanted rainfall and an irrigation system made
waterlogging of the soil possible when required. The monoliths were
collected in glass-fibre cylinders 80 cm diameter and 135 cm deep, as
described by Belford (1979). Representative crop growth was ensured by
surrounding the cropped lysimeters with a similar guard crop (Van Bavel
1961).

Clayey soils occupy about 45% of the cereal growing area of England and
Wales, and about 60% of drainage work is to alleviate surface water problems
associated with the low hydraulic conductivity in such soils. Ground water
problems account for about 30% of drainage work. We used two soil types, a
clay (an Aeric haplaquept), and a sandy loam (an Aquentic haplorthod)
representing the two main types of drainage problems, with 32 monoliths of
each to allow a reasonable number of experimental treatments with adequate
replication. Details of the lysimeter system are given by Cannell et al.
(1980a).

In Britain, the rainfall (annual mean total about 650 mm in the main crop
growing areas) is on average uniformly distributed throughout the year, and

---

Authors from: Agricultural Research Council Letcombe Laboratory, Wantage,
OX12 9JT, United Kingdom.

therefore waterlogging is most likely in winter and early spring when precipitation exceeds evapo-transpiration. Short periods of waterlogging also can occur in low conductivity soils due to summer thunderstorms. The areas of autumn (fall) sown crops are: wheat 1.45, barley 0.78, and oil-seed rape 0.13 million hectares respectively (Ministry of Agriculture, Fisheries and Food, 1981). Accordingly most attention in our research program has been given to winter wheat. Of spring crops grown on soils likely to be waterlogged, peas is the most important crop, occupying 80,000 ha.

In the absence of other information on waterlogging of crops in Britain, the experimental treatments have usually been selected as the most extreme likely to occur in field conditions, and often involved waterlogging to the soil surface. Many clayey soils have a low drainable porosity, and when at field capacity require only 30-40 mm of rain to become waterlogged in the absence of effective drainage. Thus the return periods for rainfall events to cause surface waterlogging in October and January are 5 years and 1 year respectively; and the likelihood of each event is independent of the other (Smith and Trafford 1976; Dennis, C.W., Field Drainage Experimental Unit, Ministry of Agriculture, Fisheries and Food, Cambridge, personal communication). The main experimental variables have been the position of the water-table in relation to the soil surface, the length of time that the water-table remained at that position, and the stage of growth of the crop during waterlogging. To define the treatments clearly the rainfall patterns were such as to raise the water-table rapidly to the selected depth. After waterlogging, the water-tables fell in response to natural drainage.

EFFECTS ON CROP GROWTH

In considering the effects of waterlogging on crop growth it is important to distinguish between the duration of waterlogging and the duration of anaerobic or near anaerobic conditions, because the effects of waterlogging are due primarily to the exclusion of oxygen from the roots (Cannell and Jackson, 1981). The time to deplete the oxygen from a waterlogged soil is influenced greatly by the biological demand by roots and soil organisms, and therefore is shorter in warmer soils, and close to the soil surface where biomass concentration is greatest. As a result the time for the oxygen concentration in the soil to decline from atmospheric to 2% (v/v) or less depended on the time of year. Typical results are given in Table 1. Thus the duration of near-anaerobic conditions is always less than the period of transient waterlogging, especially in cool winter months.

Table 1. Effect of Temperature on Depletion of Soil Oxygen at 20 cm Depth in a Waterlogged Clay Soil

	October	Jan/Feb	April	May
Mean soil temperature ($^{\circ}$C)	9.3	3.5	10.9	11.9
Days for oxygen at 20 cm depth to become less than 2% (v/v)	4[a]	13[a]	3.5[a]	5[b]

[a] oxygen measured with a membrane covered polarographic electrode (Blackwell 1982).
[b] oxygen measured by gas chromatography (Hall 1979).

Winter wheat

The effects of waterlogging have been examined at all the main stages of growth when a high water-table could arise in the field. The duration of waterlogging at different times was chosen as the maximum likely in practice, ranging from 16 days in October, 42 days in mid-winter and 6 days in May. Six experiments of this type have been made, and the effects on grain yield are summarized in Table 2; the 42 day mid-winter treatment was common to all the experiments. Individual waterlogging events have depressed yield by about 10% on average, with the smallest effects in years when grain yields were light (3-4 t ha^{-1}), and the greatest losses from waterlogging when yields were heavy (8-11 t ha^{-1}).

Table 2. Percentage Loss in Grain Yield of Winter Wheat After Waterlogging to the Soil Surface at Different Stages of Growth

Growth stage[a]	Pre-emergence	Tillering	Stem elongation	Flag leaf emerging	Grain yield of freely-drained
	(06)	(22)	(30)	(38)	
Month	October	Jan/Feb	April	May	control
Duration (days)	6	42	21-25	6	(t ha^{-1})
Sandy loam					
1974-5		- 8			7.6
1975-6		- 6		-5	4.0
1977-8	-16	- 7	-11		11.2
Clay					
1975-6		- 2		-6	3.8
1977-8	-18	-16	-13		9.3
1979-80	- 4[b]	-12	- 6		8.9
Mean	-13	- 9	-10	-5	7.5

[a]defined after Zadoks et al. (1974).
[b]5 days duration.

Waterlogging after germination but before emergence was the most critical stage of growth for winter wheat, with six day treatments depressing grain yield by about 17% (Table 2). However the crop was very sensitive to small changes in the duration of pre-emergence waterlogging. Few plants survived 6 days, and 10 days or more completely killed the crop at this stage (Fig. 1). Pre-emergence waterlogging for less than 4 days in Britain would probably not depress grain yield, as with usual seed rates an adequate number of plants (at least 150 m^{-2}) would survive for normal growth and yield. However, if the duration of waterlogging exceeds 4 days the plant population is depressed below a number that can be compensated for by increased tillering and in other components, and yield may be less (Table 2, 1979-80 experiment). In the three experiments listed in Table 2, the plant populations were reduced from around 350 m^{-2} in the freely-drained control to 11-42 m^{-2} after pre-emergence waterlogging. However a larger proportion of tillers survived to produce ears (Fig. 2), and the ears had more and heavier grains at harvest.

When wheat was waterlogged after emergence, plant population was not affected (provided the crop was not completely inundated). Visual symptoms of waterlogging damage were confined to more pronounced chlorosis and senescence of older leaves, and a reduction in the number of tillers. These effects were most pronounced during early vegetative growth, and with late sowing, but became less pronounced the later that waterlogging occurred during crop development. With waterlogging treatments applied during vegetative growth, i.e. between October and March, yield losses were predominently associated with fewer surviving ears (Fig. 2). When the crop was waterlogged during the grain formation period from stem elongation, i.e. April-May, then ear numbers were not affected but yield losses were due to fewer and sometimes lighter grains per ear (Cannell et al. 1980a).

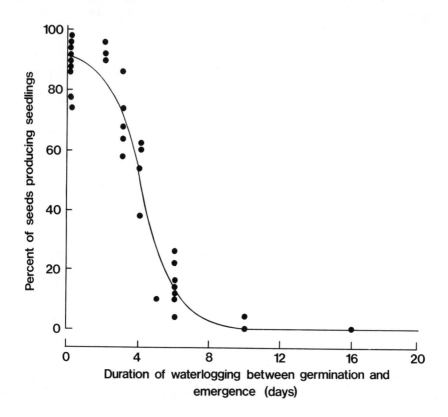

Fig. 1   Effect of Pre-emergence Waterlogging on Establishment of Winter
Wheat. Curve fitted to results from Cannell et al. (1980b),
Belford and Cannell (1981) and Thomson and Cannell (1982).

Repeated waterlogging had an additive effect on yield loss in the clay
soil. When single waterlogging events depressed yield by 4 to 12% (Table 2),
after any two of the waterlogging treatments yield was 20% less and reduced
by 30% after waterlogging on all three occasions (Cannell et al. 1981b). A
similar   trend,   but   with   smaller   losses,   occurred   after   multiple
waterloggings of winter wheat on a sandy loam (Belford 1981).

In heavy soils in Britain periods of surface waterlogging are accompanied by
longer periods of high water-table close to the soil surface, especially in
winter.   In undrained clay soils the winter water-table can fluctuate around
20 cm below the surface from December to April, the duration depending on
the winter rainfall.   Therefore in two experiments we have investigated the
effect of maintaining different depths of water-tables between December and
the end of March on winter wheat.   In an experiment in 1979-80 when the
winter water-table was at or above 20 cm, yields were about 17% less than
with the deepest (90 cm) water-table (Table 3).   In another experiment, in
1977-78, a late sown crop of winter wheat with a 20 cm water-table tillered
less (Fig. 2) and gave 18% less grain than with a 50 cm water-table (Cannell
et al. 1980b).   In 1979-80 there was little difference between yields with
50 and 90 cm winter water-tables (Table 3), suggesting that there is little
to be gained from attempting to drain clayey subsoils of very low hydraulic
conductivity.   These results show the need to lower water-tables for winter

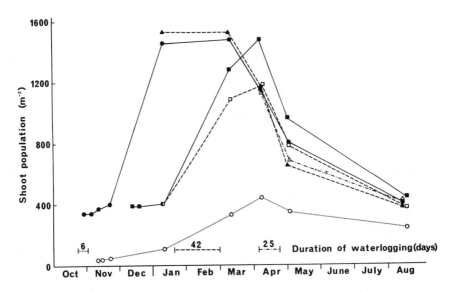

Fig. 2 Effects of Waterlogging on Shoot Populations of Winter Wheat.
Treatments sown on 26 October were: ● , freely-drained, ○, ▲ and △
waterlogged at growth stages 06, 22 and 30 respectively (defined in
Table 2); treatments sown on 29 Nov,■ and □ had water-tables at 50 and
20 cm below soil surface from Dec 10 for 160 days. (Cannell et al.
1980b).

wheat to about 50 cm. In the field, mole drains at 50-60 cm depth and 2 m
apart draining into widely spaced gravel covered pipes will lower the water-
table to about the depth of moling (Trafford and Rycroft, 1973; Goss et al.
1981). In field experiments on clayey soils in Britain where mole drainage
has been used grain yields up to about 20% greater than on undrained land
have been found (Trafford and Oliphant 1977; Armstrong 1978; Ellis et al.
1981).

Table 3. Effect of Different Depths of Winter Water-table from December
to March in a Clay Soil on the Grain Yield of Winter Wheat.

Depth of water-table (cm)	Grain yield (t ha^{-1} at 85% DM)
0	7.18
10	7.04
20	7.79
50	8.48
90	8.57
LSD	0.68

Winter barley

The area of this crop in the United Kingdom has increased greatly,
especially in the last 3 years. In more limited experimentation the effects
of pre-emergence waterlogging, and the effects of waterlogging at later
stages of growth on vegetative growth and yield have been similar to winter
wheat (Cannell et al. 1981a).

Winter oilseed rape

On the sandy loam brief periods of waterlogging (10 days) in December, January or May resulted in slightly shorter plants but hardly affected yield. Waterlogging in December/January for 6 weeks when the mean soil temperature at 20 cm depth was $2^{\circ}C$, slightly restricted growth, but yield was unaffected. Waterlogging for a similar duration between January and March (mean 20 cm soil temperature $6^{\circ}C$) halved plant size, but compensatory growth later in the season limited yield losses of seed and oil to 14 and 17%, respectively (Cannell and Belford 1980).

Peas

Spring peas are very sensitive to waterlogging. Waterlogging for more than three days after germination but before seedling emergence decreased plant populations and yield on the sandy loam. Peas, in contrast to wheat, did not have compensatory growth to offset the early effects on populations (Belford and Thomson 1980). Waterlogging to the soil surface at the 1-2, 3-4 and 6-7 leaf stages (the latter being just before flowering) gave average losses of yield of seeds (peas) at the harvesting stage for freezing of 6, 15 and 42%, respectively (Belford et al. 1980). The yield of haulm was more affected than the yield of seeds. The severity of the effect depended on the duration of the treatment and the height of the water-table. Yield was little affected following either two days waterlogging to the surface or five days with the water-table at 50 cm below the soil surface at the 6-7 leaf stage (the most sensitive stage after emergence).

CONCLUSIONS

The results indicate that, except for short periods, water-tables for the main crops in the United Kingdom should be maintained at about 50 cm or more below the soil surface. When the water-table rises nearer to the soil surface, drainage systems should be designed to remove this excess water within 4 or 5 days for winter cereals to avoid pre-emergence waterlogging, and within 2 days where peas are grown. These results should be evaluated in relation to other drainage design criteria (Trafford 1975). If the risk of waterlogging of winter cereals before emergence can be avoided by earlier sowing, or the effects lessened by heavier seed rates, wider drain spacing would be adequate to avoid direct effects of waterlogging on crop growth, because the rate of decline of the water-table would be less important.

However drainage can have both direct benefits on crop growth and indirect benefits on soil conditions. Drainage for winter cereals may be at least as important to improve trafficability for timely application of herbicides and fertilizer. For clayey soils Steinhardt and Trafford (1974) and Bailey (1979) concluded that it is advizable to maintain the water-table about 50 cm below the surface to minimize sinkage, compaction and structural damage. With clayey soils lowering of the water-table to a suitable depth to give both the direct and indirect benefits can be achieved economically by closely spaced mole drains draining into a piped system.

ACKNOWLEDGEMENT

We are most grateful to Mr B.D. Trafford who when Head of the Field Drainage Experimental Unit of the Ministry of Agriculture, Fisheries and Food at Cambridge, was closely involved in planning the experimental facilities and the early experiments; his successors Mr A.D. Bailey and Mr S. Le Grice and colleague Mr C.W. Dennis have continued to give helpful advice.

REFERENCES

1. Armstrong, A.C. 1978. The effect of drainage treatments on cereal yields: results from experiments on clay lands. J. Agric. Sci., Camb. 91: 229-235.

2. Bailey, A.D. 1979. Benefits from drainage of clay soils. Paper No. 79-2549 ASAE Winter Meeting, New Orleans. 24 pp.

3. Belford, R.K. 1979. Collection and evaluation of large monoliths for soil and crop studies. J. Soil Sci. 30: 363-373.

4. Belford, R.K. 1981. Response of winter wheat to prolonged waterlogging under outdoor conditions. J. Agric. Sci., Camb., 97: 557-568.

5. Belford, R.K., R.Q. Cannell, R.J. Thomson and C.W. Dennis. 1980. Effects of waterlogging at different stages of development on the growth and yield of peas (Pisum sativum L.). J. Sci. Food Agric. 31: 857-869.

6. Belford, R.K. and R.Q. Cannell. 1981. Effect of a calcium peroxide seed coating on the establishment of winter wheat subjected to waterlogging. Agric. Res. Council Letcombe Lab. Ann. Rep. 1980: 39-40.

7. Belford, R.K. and R.J. Thomson. 1980. Effect of waterlogging before seedling emergence on growth and yield of peas. Agric. Res. Council Letcombe Lab. Ann. Rep. 1979: 27-28.

8. Blackwell, P.S. 1982. Measurements of aeration in waterlogged soils: some improvements of techniques and their application to experiments using lysimeters. J. Soil. Sci. (in press).

9. Bouwer, H. 1974. Developing drainage design criteria. p. 67-90. In: J. Van Schilfgaarde (ed.) Drainage for agriculture. Am. Soc. Agron., Madison, Wl.

10. Cannell, R.Q. and R.K. Belford. 1980. Effects of waterlogging at different stages of development on the growth and yield of winter oilseed rape (Brassica napus L.). J. Sci. Food Agric. 31: 963-965.

11. Cannell, R.Q., R.K. Belford, P.S. Blackwell and R.J. Thomson. 1982. Effects of different depths of water-table on the growth and yield of wheat on a clay soil. Agric. Res. Council Letcombe Lab. Ann. Rep. 1981. (in press).

12. Cannell, R.Q., R.K. Belford, K. Gales and C.W. Dennis. 1980a. A lysimeter system used to study the effect of transient waterlogging on crop growth and yield. J. Sci. Food Agric. 31: 105-116.

13. Cannell, R.Q., R.K. Belford, K. Gales, C.W. Dennis and R.D. Prew. 1980b. Effects of waterlogging at different stages of development on the growth and yield of winter wheat. J. Sci. Food Agric. 31: 117-132.

14. Cannell, R.Q., R.K. Belford and R.J. Thomson. 1981a. Effect of waterlogging at different stages of development on the growth and yield of winter barley. Agric. Res. Council Letcombe Lab. Ann. Rep. 1980: 33-34.

15. Cannell, R.Q., R.K. Belford and R.J. Thomson. 1981b. Effects of single and multiple waterloggings on the growth and yield of winter wheat on a clay soil. Agric. Res. Council Letcombe Lab. Ann. Rep. 1981: 34-35.

16. Cannell, R.Q. and M.B. Jackson. 1981. Alleviating aeration stresses. p. 141-192. In: Modifying the root environment to reduce crop stress G.F. Arkin and H.M. Taylor (eds.). ASAE St. Joseph, Michigan.

17. Ellis, F.B., D.G. Christian and R. Crees. 1981. Long-term drainage/cultivation experiment: agronomic aspects and yield. Agric. Res. Council Letcombe Lab. Ann. Rep. 1980: 16-17.

18. Goss, M.J., K.R. Howse and J.M. Vaughan-Williams. 1981. Long-term drainage/cultivation experiment: effect of mole drainage on soil water regimes and water extraction. Agric. Res. Council Letcombe Lab. Ann. Rep. 1980: 17.

19. Hall, K.C. 1979. A gas chromatographic method for the determination of oxygen dissolved in water using an electron capture detector. J. Chromatog. Sci., 16: 311-313.

20. Hiler, E.A. 1976. Drainage requirements of crops. p. 127-129. In: Proc. 3rd Nat. Drainage Conf., ASAE, St. Joseph, Michigan.

21. Ministry of Agriculture, Fisheries and Food. 1981. Agricultural statistics - England, 2. MAFF, London.

22. Raadsma, S. 1974. Current drainage practices in flat areas of humid regions of Europe. p. 115-143. In: J. van Schilfgaarde (ed.) Drainage for agriculture. Am. Soc. Agron., Madison, Wl.

23. Smith, L.P. and B.D. Trafford. 1976. Climate and Drainage. Technical Bulletin No. 34. Min. Agric. Fish. Food, London.

24. Steinhardt, R. and B.D. Trafford. 1974. Some effects of sub-surface drainage and ploughing on the structure and compactability of a clay soil. J. Soil Sci. 25: 138-152.

25. Thomson, R.J. and R.Q. Cannell. 1982. Effects of a calcium peroxide seed coating on the establishment of winter wheat waterlogged before emergence. Agric. Res. Council Letcombe Lab. Ann. Rep. 1981. (in press).

26. Trafford, B.D. 1975. Improving the design of practical field drainage. Soil Sci. 119: 334-338.

27. Trafford, B.D. and J.M. Oliphant. 1977. The effect of different drainage systems on soil conditions and crop yield on a heavy clay soil. Expl. Husb. 32: 75-85.

28. Trafford, B.D. and D.W. Rycroft. 1973. Observations on the soil-water regimes in a drained clay soil. J. Soil Sci. 24: 380-391.

29. Van Bavel, C.H.M. 1961. Lysimetric measurements of evapotranspiration rates in the Eastern United States. Proc. Soil Sci. Soc. Am. 25: 138-141.

30. Zadoks, J.C., T.T. Chang and C.F. Konzak. 1974. A decimal code for the growth stages of cereals. Weed Res. 14: 415-421.

BENEFITS FROM THE DRAINAGE OF HEAVY IRRIGATED SOILS

WILLIAM R. JOHNSTON, BYRON C. STEINERT, AND CRAIG M. STROH[1]
MEMBER
ASAE

Many irrigated areas of the world have subsurface agricultural drainage problems. A typical area with such a problem is the semiarid San Joaquin Valley in California.

There are 3,200,000 hectares (8,000,000 acres) of irrigated land in the Valley, of which 160,000 hectares (400,000 acres) currently have a severe saline drainage problem. Eventually, more than 400,000 hectares (1,000,000 acres) will also have a drainage problem that must be overcome to maintain optimum production on the land (Johnston and Beck 1978). Most of the soils which have or will have a drainage problem are stratified heavy clay, silty clay or clay loam soil types (Storie and Weir 1953), and are located on the west side of the San Joaquin Valley, but near the trough of the Valley. Figure 1 shows the Drainage problem area and the location of Westlands Water District in the San Joaquin Valley, California.

THE PROBLEM

The stratified clay soil layers cause a perched saline water table to develop. Once this perched water table has risen to the root zone it prevents leaching of native salts and salts in the applied irrigation water.

An increase in irrigation over several decades throughout the Valley has caused an increase in the extent of the drainage problem, i.e., the area with a saline groundwater problem. The perched water tables contain on the average 10,000 to 15,000 milligrams per liter - Total Dissolved Solids (mg/l-TDS).

When a saline groundwater table rises into the crop root zone and is not corrected with proper drainage, a decline in crop yields will result and a change in cropping patterns to more salt-tolerant crops may be necessary. This is due both to the high water table and the resultant increased salinity in the crop root zone (Ayers and Westcot 1976). Christopher and TeKrony (1982) estimated the importance of water table and salinity control on crop production. However, they concluded that their analysis was not adaquately supported by desired experience and research. Once the saline water table rises to within 1.5 meters (5 feet) of the ground surface, crop yields can be reduced by as much as 20 percent in a period of a few years unless on-farm drainage is provided (Johnston 1978). Profits from farming can diminish quickly under such conditions, leading to a severe or total loss in the agricultural value of affected land.

1/ The authors are: William R. Johnston, Assistant Manager-Chief of Operations, and Byron C. Steinert, Senior Engineer, Westlands Water District, P. O. Box 6056, Fresno, CA 93703, and Craig M. Stroh, Economist, U.S. Bureau of Reclamation, 2800 Cottage Way, Sacramento, CA 95825.

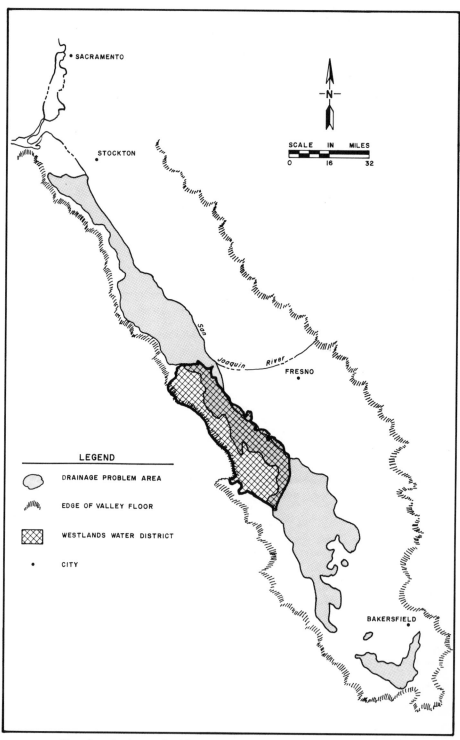

SACRAMENTO

STOCKTON

SCALE   IN   MILES

0   16   32

San   Joaquin   River

FRESNO

LEGEND

DRAINAGE PROBLEM AREA

EDGE OF VALLEY FLOOR

WESTLANDS WATER DISTRICT

CITY

BAKERSFIELD

Fig. 1   Drainage Problem Area and Westlands Water District
          in the San Joaquin Valley of California

West Side San Joaquin Valley

A May 1982 survey of water districts on the west side of the San Joaquin
Valley reveals the extent and severity of the region's drainage problems.
Of the almost 486,000 hectares (1,200,000 acres) covered by the survey,
more than 160,000 hectares (400,000 acres) have a perched water table
within 3.0 meters (10 feet) of the ground surface. More than 30,000 of
these hectares (75,000 acres) now have on-farm drainage systems. In
anticipation of the growth of the problem, the survey projects that by
1995 more than 81,000 hectares (200,000 acres) will have subsurface
drainage systems installed.

Westlands Water District

Westlands Water District, a political subdivision of California, is part
of the west side of the San Joaquin Valley, lying between the foothills of
the Coast Range on the west and the Valley trough on the east. Westlands,
covering approximately one-half of the 486,000 hectares (1,200,000 acres)
of west side farmland surveyed, is the largest water district in California
and the United States in terms of irrigated land. It is estimated that about
one-half of the District area ultimately will need subsurface agricultural
drainage facilities. As of October 1981, over 107,000 hectares (264,000
acres) of land within the District had a perched water table less than 6.1
meters (20 feet) below the ground surface. Table 1 shows the growth of the
areas affected by the perched water table between 1967 and 1981.

Table 1: Increase in Area with a Perched Water Table in Westlands Water
          District between October 1967 and October 1981.

Depth to water		Oct. 1967		Oct. 1981		14-Year Difference	
(Meters)	(Feet)	(Hectares)	(Acres)	(Hectares)	(Acres)	(Hectares)	(Acres)
0-1.5	0-5	4,980	12,300	7,570	18,700[a]	+ 2,590	+ 6,400
1.5-3.0	5-10	23,070	57,000	55,040	136,000	+31,970	+ 79,000
3.0-6.1	10-20	36,180	89,400	44,560	110,100	+ 8,380	+ 20,700
		64,230	158,700	107,170	264,800	+42,940	+106,100

[a]The October 1981 area with a 0-1.5 meter (0-5 foot) water table is
probably larger than the size indicated. The groundwater levels
measured in wells installed in 18,200 hectares (45,000 acres) of
District land served by a drainage collector system completed in
1980 are influenced by the 2.1 to 3.3 meter (7-11 foot) deep collector drains. After the drains were installed, most of the wells
ended up located too close to the drains to measure water table
levels beyond the influence of the drains. Additional wells will
have to be installed to monitor accurately the actual water table
levels.

DRAINAGE BENEFITS

Drainage benefits are defined as the net farm income maintained by an on-
farm drainage system less drainage costs. Unfortunately, little actual
data measuring the impact of saline perched water on agricultural production
are available for semiarid areas with a high saline water table. This
lack of data extends both to reductions in potential production when the
saline water table rises into the crop root zone and to the restoration in
agricultural output from providing on-farm drainage facilities in drainage
problem areas.

Johnston and Beck (1978) showed that the weighted annual net farm income in the District averaged about $370 per hectare ($150 per acre). An on-farm drainage system can preserve only a portion of the net farm income lost due to inadequate drainage. This is because the cost of installing and maintaining the on-farm drainage system must be deducted from whatever income is gained through drainage.

Another factor to be considered is that there is a lag between recognition of the drainage problem and restoration of income (productivity). In general, three lags can be identified. The first is the time between the beginning of the production decline and the recognition that a problem has developed. The second extends to the actual installation of on-farm drain-age facilities. The third lasts until productivity is restored. The income lost during these lags is irrecoverable.

In addition to preservation of income, indirect benefits also are realized by local and regional economies through adequate drainage. The agricul-tural economic base is maintained and employment levels are sustained. Farm suppliers and manufacturers of farm-related equipment will benefit from sustained demand for their products while their workers will enjoy higher wages and employment levels. Farm product processors will benefit from preparing and handling a greater amount of food and fiber for ship-ment to merchant and consumer.

A four-year coordinated State of California - U.S. Bureau of Reclamation drainage study report (San Joaquin Valley Interagency Drainage Program 1979) indicated that the output value of typical crops grown in the District drainage problem area has a multiplier effect of about 2.3 for the San Joaquin Valley and about 3.0 for California as a whole. This means that each dollar of lost output in the District due to the drainage problem will cause a $2.30 loss in total Valley income and a $3.00 loss in state-wide income.

Crop Production

Data showing cropping patterns, crop yields, unit prices, and groundwater conditions are unavailable for the west side of the Valley as a whole. Each year, however, Westlands Water District compiles a Crop Production Report for the U. S. Bureau of Reclamation showing the area, the average yield, the average unit price, and total value of each crop produced in the District. Crop data from 1979, 1980, and 1981 have been compiled for three different areas of the District, the area having less than a 1.5 - meter (5-foot) water table, the area with a water table between 1.5 and 3.0 meters (5-10 feet), and the remaining areas of the District where the perched water table is assumed to have no impact on crop production. The total gross value of production (gross income) for a composite hectare (acre) of the different crops produced in the three respective water table areas of the District for 1979, 1980, and 1981 is presented in Table 2. The reductions in gross income due to perched saline water table, as shown in Table 2, are conservative estimates of the actual losses because the same crop yields are used for the different areas, whereas crop yields in the areas with higher saline water tables are expected to be lower. It is difficult, however, to confirm variations in crop yields among farms in different areas because of divergent farming practices.

Table 2: The Gross Income Value of Production of Crops of a Composite
Hectare (Acre) in Different Water Table Areas of Westlands Water
District.

Area[a]	1979			1980			1981		
Water Table	$/ha	$/ac	%[b]	$/ha	$/ac	%[b]	$/ha	$/ac	%[b]
< 1.5m	1386	561	70	1809	732	79	1710	692	77
1.5-3.0m	1764	714	89	2105	852	91	1977	800	89
> 3.0m	1984	803	100	2303	932	100	2214	896	100

[a] Area of District with perched water table as indicated.

[b] Percent of value in the area with water table greater than 3.0
meters (10 feet) deep.

RESULTS AND DISCUSSION

As seen from the data in Table 1 both the extent and severity of the District's drainage problems have increased significantly since imported
irrigation water was first applied to District land in the late 1960's.
The area with a water table within 3 meters (10 feet) of the ground surface
has more than doubled in this time period; the area with a saline water
table within 1.5 meters (5 feet) of the surface (which can be taken as an
index of the problem's severity) has increased by more than 50 percent.
As shown in Table 2, in each of the three most recent crop years, the
gross value of production on the most severely affected land has ranged
from only 70 percent to 77 percent of that on land without a drainage
problem. The first thing a farmer is likely to do in reaction to a developing drainage problem if he does not install drainage facilities is change
his cropping pattern to more salt tolerant, lower value crops where possible.
While data in Table 2 show the differences in gross crop income resulting
from cropping pattern changes, they do not show production costs or profit.
Without drainage, the yields of even the more salt tolerant crops will
decline due to salt buildup. This will result in the farmer eventually
going out of business. Farming in a large area of the District may become
unprofitable in the near future.

In some areas of the western United States, and in many humid areas of the
world, nonsaline subsurface drainage can be disposed of through reuse for
irrigation or discharged without any problem in local watercourses. The
high salinity of the District's drainage water has prevented its agricultural reuse or disposal in the nearby San Joaquin River. The District has
been working with the U. S. Bureau of Reclamation in order to get the San
Luis Drain completed. This Drain would convey saline drainage from the
District to the Sacramento-San Joaquin Delta estuary for safe disposal.

As noted earlier, the weighted annual net farm income with drainage is
approximately $370 per hectare ($150 per acre). The estimated total annual
cost to the grower of on-farm drainage facilities and a system to dispose
of the effluent safely in the Sacramento-San Joaquin Delta was estimated
to be some $99 per hectare ($40 per acre) (San Joaquin Valley Interagency
Drainage Program 1979). Comparison of these benefits and costs clearly
indicates that it is preferable to install subsurface drainage rather than
allow land to go out of production.

The individual grower, faced with a developing drainage problem, may not react promptly and make a decision to invest in an on-farm drainage system. First of all, as previously noted, he may not even recognize the nature and/or severity of the problem. Initial productivity declines, though real, may be relatively minor and go unnoticed for some time. Unfortunately, when the problem becomes severe and yield losses significant, difficulties may be encountered in the physical installation of a drainage system because of a high water table and soil saturation (Johnston 1978). The associated income loss is irrecoverable and should prompt growers in potential drainage problem areas to monitor carefully crop yields and soil conditions.

In addition, a recognized, needed investment in drainage is subject to more than a purely economic benefit/cost comparison. It also depends on the grower's financial situation. For example, two growers facing identical drainage problems may react very differently because one has a high degree of equity in his farm while the other's farm is highly mortgaged. The first can use his equity, or even a cash reserve, to finance a drainage system while the second lacks loan security. Furthermore, the second grower would have larger fixed debt payments to make, thereby reducing his ability to service additional debt.

Estimates of economic losses due to San Joaquin Valley drainage problems are difficult to make because of a lack of accurate data on depths to water tables, salinity, changes in cropping patterns, and yield reductions. Some estimates, however, can be made based on drainage problems experienced in Westlands Water District. It should be noted that these estimates measure economic impacts of drainage problems occurring in only a portion of the Valley's drainage problem area. Based on data in Tables 1 and 2 for areas with a greater than 3 meter (10 foot) water table in Westlands Water District, it is estimated that crop production losses due to a perched saline water table amounted to nearly $17,000,000 in 1981 and averaged more than $16,000,000 yearly for the period 1979-1981. This would translate to an annual income loss of almost $37,000,000 for the Valley and $50,000,000 Statewide.

SUMMARY

Stratified clay soil layers occurring along the west side of the San Joaquin Valley have caused a perched saline water table to develop. When the water table rises into the crop root zone, a decline in crop yields and/or a change to more salt tolerant crops will result. Under these conditions, farm profits can decline rapidly.

On-farm subsurface drainage can hold the level of the water table below the crop root zone so that productivity losses and/or changes in cropping patterns can be prevented or reversed. Economic benefits attributable to a drainage system are measured by the net farm income maintained. The annual net on-farm income maintained by drainage in Westlands Water District is estimated at $370 per hectare ($150 per acre) with the grower's costs for a complete system--from on-farm facilities to ultimate disposal of the saline effluent--estimated at some $99 per hectare ($40 per acre). Clearly, investment in drainage--particularly under circumstances where drainage effluent is too salty to be recirculated in the irrigation supply or discharged to a nearby watercourse--is profitable for the grower facing a high water table and declining crop yields.

The beneficial economic impacts of drainage are felt beyond the confines of the farm and the grower's profits. Avoiding Westlands' $17,000,000 drainage related production loss for 1981 would have meant about $37,000,000 more income in the entire San Joaquin Valley and an additional $13,000,000 more income in the rest of the State. All citizens would benefit, directly or indirectly, from investment in agricultural drainage.

REFERENCES

1. Ayers, R. S. and D. W. Westcot. 1976. Water Quality for Agriculture. FAO Irrig. & Drain. Paper 29. Rome.

2. Christopher, J. N. and R. G. TeKrony. 1982. Benefits Related to Water Table and Salinity Control. Amer. Soc. Agri. Engr. Paper No. 82-2077.

3. Johnston, William R. 1978. Drainage Installation Problems in the San Joaquin Valley, California, U.S.A., Proc. Inter. Drainage Workshop, Wageningen. The Netherlands. pp. 603-619. J. Wesseling, Editor.

4. Johnston, William R. and Louis A. Beck. 1978. Financing Regional Drainage Facilities Under 1978 Economic Conditions. Amer. Soc. Agri. Engr. Paper No. 78-2535.

5. San Joaquin Valley Interagency Drainage Program, Final Report 1979. Agricultural Drainage and Salt Management in the San Joaquin Valley. U.S. Bureau of Reclamation California Department of Water Resources and California State Water Resources Control Board.

6. Storie, R. Earl and Walter W. Weir. 1953. Generalized Soil Map of California. Calif. Agric. Exp. Sta. Manual 6.